T0212113

Lecture Notes in Computer Science 14622

Founding Editors

Gerhard Goos
Juris Hartmanis

The series Lecture Notes in Computer Science (LNCS), including its subseries Lecture Notes in Artificial Intelligence (LNAI) and Lecture Notes in Bioinformatics (LNBI), has established itself as a medium for the publication of new developments in computer science and information technology research, teaching, and education.

LNCS enjoys close cooperation with the computer science R & D community, the series counts many renowned academics among its volume editors and paper authors, and collaborates with prestigious societies. Its mission is to serve this international community by providing an invaluable service, mainly focused on the publication of conference and workshop proceedings and postproceedings. LNCS commenced publication in 1973.

Tiago Dias · Paola Busia

Editors

Design and Architectures for Signal and Image Processing

17th International Workshop, DASIP 2024
Munich, Germany, January 17–19, 2024
Proceedings

 Springer

Editors
Tiago Dias
Instituto Superior de Engenharia de Lisboa
Lisbon, Portugal

Paola Busia
Università degli Studi di Cagliari
Cagliari, Italy

ISSN 0302-9743   ISSN 1611-3349 (electronic)
Lecture Notes in Computer Science
ISBN 978-3-031-62873-3   ISBN 978-3-031-62874-0 (eBook)
https://doi.org/10.1007/978-3-031-62874-0

This Springer imprint is published by the registered company Springer Nature Switzerland AG
The registered company address is: Gewerbestrasse 11, 6330 Cham, Switzerland

If disposing of this product, please recycle the paper.

Preface

This volume contains the papers presented at the 2024 Workshop on Design and Architectures for Signal and Image Processing (DASIP 2024), which was held jointly with the 19th HiPEAC Conference in Munich, Germany, on January 17–19, 2024. The workshop provided an inspiring international forum for the latest innovations and developments in the fields of leading signal, image, and video processing and machine learning in custom embedded, edge, and cloud computing architectures and systems.

DASIP is a long-running annual workshop that was organized for the first time in 2007 in Grenoble, France, and since then it has been held in several locations in Europe and Canada, which include: Toulouse, France (DASIP 2023); Budapest, Hungary (DASIP 2022); Montréal, Canada (DASIP 2019); Porto, Portugal (DASIP 2018); Dresden, Germany (DASIP 2017); Rennes, France (DASIP 2016); Krakow, Poland (DASIP 2015); Madrid, Spain (DASIP 2014); Cagliari, Italy (DASIP 2013); Karlsruhe, Germany (DASIP 2012); Tampere, Finland (DASIP 2011); Edinburgh, UK (DASIP 2010); Sophia Antipolis, France (DASIP 2009); and Brussels, Belgium (DASIP 2008).

For this 17th edition of DASIP, 21 papers were submitted from 12 countries around the world. Each contributed paper underwent a rigorous double-blind peer-review process during which it was reviewed by at least three reviewers who were drawn from a large pool of the Technical Program Committee members and some external reviewers. As a result, 9 high-quality papers were accepted for oral presentation at the workshop and published in these proceedings.

The success of DASIP depends on the contributions of many individuals and organizations. With that in mind, we thank all authors who submitted their work to the conference. We also wish to offer our sincere thanks to the members of the Technical Program Committee for their very detailed reviews, and to the session chairs for contributing to the success of DASIP 2024. We further extend our appreciation to the members of the Steering Committee for their support.

We address special thanks to David González-Arjona, from GMV Aerospace & Defence (Spain), for presenting a deeply inspiring keynote during the event, and to Andrea Pinna, from Sorbonne University (France), Marek Gorgon, from AGH University of Science and Technology (Poland), and Paolo Meloni, from University of Cagliari (Italy), for making the panel discussion on the future challenges of embedded AI for image and signal processing another memorable and valuable session at DASIP 2024.

January 2024

Tiago Dias
Paola Busia

Organization

General Chairs

Tiago M. Dias   ISEL – IPL/INESC-ID, Lisbon, Portugal
Paola Busia   University of Cagliari, Italy

Steering Committee

João M. P. Cardoso	University of Porto, Portugal
Miguel Chavarrías	Universidad Politécnica de Madrid, Spain
Jean-Pierre David	Polytechnique Montréal, Canada
Karol Desnos	IETR, INSA Rennes, France
Diana Göhringer	Technical University of Dresden, Germany
Marek Gorgon	AGH University of Science and Technology, Poland
Michael Huebner	Brandenburg University of Technology, Germany
Tomasz Kryjak	AGH University of Science and Technology, Poland
Pierre Langlois	Polytechnique Montréal, Canada
Paolo Meloni	University of Cagliari, Italy
Andrea Pinna	Sorbonne University, France
Sebastien Pillement	Polytech Nantes, France
Sergio Pertuz	Technical University of Dresden, Germany
Alfonso Rodriguez	Universidad Politécnica de Madrid, Spain

Program Committee

Gabriel Caffarena	CEU San Pablo University, Spain
João Canas Ferreira	University of Porto, Portugal
João M. P. Cardoso	University of Porto, Portugal
Miguel Chavarrías	Universidad Politécnica de Madrid, Spain
Daniel Chillet	IRISA/ENSSAT, University of Rennes 1, France
Christopher Claus	Robert Bosch GmbH, Germany
Martin Danek	Daiteq s.r.o., Czechia
Jean-Pierre David	Polytechnique Montréal, Canada
Eduardo de la Torre	Universidad Politécnica de Madrid, Spain

Karol Desnos	IETR, INSA Rennes, France
Milos Drutarovsky	Technical University of Kosice, Slovak Republic
Diana Göhringer	Technical University of Dresden, Germany
Guy Gogniat	Université de Bretagne Sud, France
Marek Gorgon	AGH University of Science and Technology, Poland
Bertrand Granado	Sorbonne University, France
Oscar Gustafsson	Linköping University, Sweden
Frank Hannig	University of Erlangen-Nurnberg, Germany
Mateusz Komorkiewicz	Aptiv, Poland
Tomasz Kryjak	AGH University of Science and Technology, Poland
Yannick Le Moullec	Tallinn University of Technology, Estonia
Gustavo M. Callico	Univ. of Las Palmas de Gran Canaria, Spain
Kevin J. M. Martin	Université de Bretagne Sud, France
Paolo Meloni	University of Cagliari, Italy
Jean-François Nezan	IETR, INSA Rennes, France
Arnaldo Oliveira	University of Aveiro, Portugal
Andrés Otero	Universidad Politécnica de Madrid, Spain
Francesca Palumbo	University of Sassari, Italy
Maxime Pelcat	INSA Rennes/IETR Laboratory, France
Sergio Pertuz	Technical University of Dresden, Germany
Christian Pilato	Politecnico di Milano, Italy
Sebastien Pillement	University of Nantes, France
Andrea Pinna	Sorbonne University, France
Jorge Portilla	Universidad Politécnica de Madrid, Spain
Nuno Roma	University of Lisbon, Portugal
Olivier Romain	ETIS, France
Ruben Salvador	CentraleSupélec, France
Dimitrios Soudris	National Technical University of Athens, Greece

Additional Reviewers

Luis Crespo	University of Lisbon, Portugal
Lester Kalms	Technical University of Dresden, Germany
Ahmed Kamaleldin	Technical University of Dresden, Germany
Panagiotis Minaidis	National Technical University of Athens, Greece
Nuno Neves	University of Lisbon, Portugal
Ariel Podlubne	Technical University of Dresden, Germany
João Vieira	University of Lisbon, Portugal

Organizational Sponsors

Invited Talk

Orbiting the Edge and Stars: Bridging the Gap Between Space Avionics and Edge Computing, Challenges, and Space Missions Needs

David González-Arjona[1,2]

[1] GMV Aerospace & Defence, Spain
dgarjona@gmv.com
[2] Universidad Autónoma de Madrid, Spain

Space missions present not only challenges but also opportunities for new high-performance computing and architectural designs. Bridging the gap between space avionics and edge computing is important to show how the challenges faced in space are relevant and interconnected with the challenges encountered for other industries edge computing. Space missions are strongly based in remote processing capabilities in autonomous manner. Communications to ground may be costly, not continuous, limited in bandwidth or with big latencies for real-time actuation needs. Earth Observation and space-based sensors are generating and consuming bigger amount of data and edge computing facilitates communication reduction and real-time executions. Exploration missions and landing operations in other solar system celestial bodies autonomously may impose high-performance solution challenges into avionics electronics that must be reliant and enduring harsh environmental conditions. Fault Tolerant hardware/software architectures and technology robust against cosmic radiation, extreme temperature gradients and mechanical vibrations may limit performance capabilities. Last but not least, machine learning applications are entering more and more in the mission's algorithms, including computing challenges to get the functional benefits desired. Continual learning possibilities will deal with uncertainties in the explored scenario but also expose avionics towards architectural challenges. These topics will be presented and discussed including examples and roadmaps for key technology for the near future in space.

Contents

Specialized Hardware Architectures
for Signal and Image Processing

A Highly Configurable Platform for Advanced PPG Analysis

Flavie Durand De Gevigney[1], Julien Heulot[1], Eric Bazin[1],
Jean-François Nezan[1], Mickaël Dardaillon[1(✉)], and Slaheddine Aridhi[2]

[1] Univ Rennes, INSA Rennes, CNRS, IETR - UMR6164, Rennes, France
`{flavie.gevigney,julien.heulot,eric.bazin,jean-francois.nezan,`
`mickael.dardaillon}@insa-rennes.fr`
[2] Sensoria Analytics, Valbonne, France
`slah@sensoriaanalytics.com`

Abstract. Growing interest in healthcare applications has made biomedical engineering one of the fastest growing disciplines in recent years. Photoplethysmography (PPG) is one of these trends since PPG sensors are embedded into many devices like smartwatches or oximeters. Due to the versatility of PPG, it is the technique of choice for the non-invasive monitoring of vital signs such as heart rate, respiratory rate, blood oxygen saturation and blood pressure. However, commercial pulse oximeters often use proprietary data acquisition and visualization techniques, making digital signal processing on raw data difficult or impossible. This paper presents a hardware platform for the exploration of new biological signal analysis based on PPG. The originality of the platform is to expand the number of wavelengths to 8 and to allow the parameterization of the acquisition by choosing the sample rate. This allows the PPG sensor to be adapted to the needs of analysis on the whole signal or perform high precision measurements. Much consideration has been given to the accurate data acquisition during sports activities.

Keywords: Photoplethysmography · PPG · Multi-wavelength · Mixed Signal · Heart Rate · SpO2

1 Introduction

Photoplethysmography (PPG) is a simple and low-cost optical technique that can be used to detect blood volume changes in the microvascular bed of tissue. It is often used non-invasively to make measurements at the skin surface (Fig. 1 [6]). As an optical technique, PPG requires a light source (LED) and a photodetector to operate. The light passes through the tissue and the photodetector detects small changes in light intensity.

When the heart pumps blood, each contraction of the heart causes an increase in the amount of blood in the capillaries in the skin's surface, resulting in more

With the support of ANR in the framework of the PIA EUR DIGISPORT project (ANR-18-EURE-0022).

light absorption. The blood then travels back to the heart through the venous network, leading to a decrease of blood volume in the surface capillaries, resulting in less light absorption. As shown on Fig. 1, the signal can be decomposed in two parts, the steady component and pulsating component. Each part contains useful information and is used to compute physiological features. To calculate certain biomarkers, it is necessary to have a precise value for the constant component (DC) and the pulsed component (AC). A particular attention should be given to the acquisition of the pulsed component because its amplitude represents only about 5% of the constant component.

PPG signals can be acquired by either transmittance or reflectance. In transmittance mode, the photons leave the LED, pass through the finger and are measured by the photodiode. In reflection mode, the skin and tissues reflect the photons, which are captured by the photodiode placed near the LED. Transmittance mode produces good signals, but can only be used on body sites of limited thickness, such as the fingertips, earlobes or nasal septum. The reflection mode does not have this problem and allows measurements to be taken on other body sites such as the wrists, chest, or forehead.

One of the fundamental principles of PPG relies on the higher sensitivity of certain optical wavelengths for blood rather than other tissue components. The Red LED ($\sim$660 nm) and Infrared (IR) LED ($\sim$940 nm) are the standard wavelengths used in pulse oximeters for the calculation of oxygen saturation. The green LED wavelength ($\sim$550 nm) is being widely used in reflection-mode applications, since tissue reflects well at this wavelength. As the evaluation of quantity and concentration using PPG is based on Beer-Lambert law, the more signals with different wavelengths are measured, the more accurate and robust the evaluation is.

This paper considers the use of PPG in two domains: medical and sports. The medical domain focuses on detecting, tracking and curing diseases, while the sports domain targets training stress tracking and performance enhancement. However, despite the different goals of the two domains, both require measurements of the same physiological features and the same dedication to precision and robustness.

To answer the challenges posed by PPG acquisition, we propose a new hardware sensor that is highly configurable. Section 2 describes the background and the necessity of having a configurable hardware for PPG. The control given to the user allows them to adapt the sensor to the experiment or even to the test subject, to ensure the best result possible. Section 3 presents the platform design and the hardware choices that were made, to achieve the necessary configurability and precision. Section 4 presents the experimental results obtain using this platform. Finally, Sect. 5 discusses the findings and presents the conclusion.

2   Background

This section presents the background of the state of the art of PPG acquisition, utilization, and limitations, highlighting the benefits of a more configurable platform with better access to raw output data. First, PPG utilization is presented to

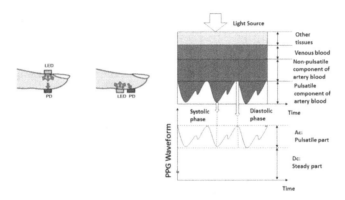

Fig. 1. Principle of the PPG (from [6]).

understand which features are important and how the quality of the output data influences the result of the biomarker calculation. Next, related work on existing platforms are detailed, highlighting the gaps in current commercial offers.

2.1 PPG Analysis

Heart Rate (HR) is the frequency of the beating of the heart. It is a good indication of the physical condition and provides valuable information, in particular in a sporting context. Heart rate monitoring for fitness is valuable for athletes to monitor their workload. The blood volume change detected by the PPG is correlated to the heart rate. The pulse rate is effectively equivalent to the heart rate and can be calculated using PPG [2,3,8].

In order to do so, different techniques can be used. Most simply, the peaks and valleys of the signal are detected and used to deduce the pulse rate [8]. A technique using the frequency domain can also be used [7]. The accuracy of the pulsating part (sampling rate and quantization) of the signal has an overarching importance for this feature.

Heart Rate Variation (HRV) quantifies the variation in time between each heartbeat. This parameter is tracked primarily in sports applications as it changes during and after exercise. The HRV feature can be used to analyse the stress the body experiences during training and to gain insight into physiological recovery after training. Information regarding the extent to which the body recovers after training provides useful data for the personalization of sports training, training loads, and recovery time. HRV monitoring of athletes is frequently applied to the prevention and diagnosis of overtraining syndrome [4]. As well as HR, HRV computation accuracy is tied to the pulsating component of the captured signal. High sample rate and high quantization accuracy of the signal are necessary for this metric.

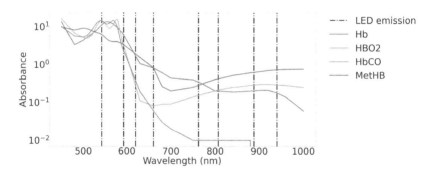

Fig. 2. Absorbance of Hemoglobin (Hb), Oxyhemoglobin (HbO2), Carboxyhemoglobin (HbCO), Methemoglobin (MetHb) and wavelength of the LEDs chosen for the platform.

Oxygenation and Peripheral Oxygen Saturation (SpO2). The blood oxygenation can be deduced from the PPG signal, as the oxygenated hemoglobin has a different absorption to the deoxygenated hemoglobin [9] when they are exposed to multiple wavelengths, as illustrated on Fig. 2. HbCO and MetHb are often neglected, as their concentration in the blood under normal circumstance is very low to non-existent. The use of absorption of light for calculation of SpO2 levels makes use of the Beer-Lambert Law. Using additional wavelengths enable the detection of those types of hemoglobin.

Breathing. The effect of breathing can be considered as noise or as another signal, depending on the PPG analysis to be performed. Indeed, respiration and synaptic nerve activity influence the final signal, making some analyses difficult to perform. However, at the same time, it is an interesting biomarker that can be extracted. Physiological effects related to respiratory activity and the cardiac cycle induce several modulations of the PPG. For this metric, accurate quantization of the signal and its sampling rate are a key of accurate computation.

2.2   Related Work

This section surveys some available devices used to extract and study PPG signals. Several technologies are categorized depending on the intended use. Whilst most of these devices allow the visualization, storage, and extraction of parameters from the PPG signal, they offer limited possibilities for additional data extraction and novel signal processing algorithms.

Platforms that allow experimental analysis should be able to make comparisons of different algorithms. Commercial solutions generally do not disclose the details of their hardware and software. From our perspective, existing PPG devices can be categorized in packaged solutions and evaluation systems.

Packaged PPG Devices. Nowadays, activity trackers and smartwatches embed PPG sensors to perform HR analysis and SpO2 measurements. The raw data measured by these devices are generally not accessible by the user; the user only has access to the compiled data (mean, max, or min) or even proprietary analysis including sleep analysis or stress level. These devices are dedicated to consumer applications and do not allow for advanced analysis required for research projects. Despite the acceptable accuracy of the displayed metrics (such as HR, HRV, SpO2, RR interval, etc.), further analysis is impossible due to inaccessibility of the raw data [5].

PPG Evaluation/Development Kits. Some chip manufacturers and companies propose kits for developing projects and applications using physiological sensors. As an example of dedicated company, BITalino proposes a handy plug and play system that allows many physiological measurements [1]. While these kinds of devices are useful to develop projects in a limited time, the performance of sensors are limited to commercial sensor performance, i.e. 500 Hz sample rate with 2 or 3 wavelengths. Moreover, they do not allow tuning of all the different building blocks needed for PPG signal acquisition and analysis.

The MAX30101 Evaluation kit (EV kit) from Maxim Integrated allows the evaluation of the MAX30101, which is an integrated pulse oximetry and heart-rate monitor integrated circuit. The MAX30101 includes internal LEDs (green, red and IR), a photodetector together with low-noise electronics including ambient light rejection and digital filtering. The MAX30101 EV kit monitors and stores the recorded PPG signals while allowing the user to modify the sampling rate (up to 3200 Hz), LED currents, and pulse widths. This EV kit board allows the user to fully exploit the MAX30101, which is a 3-wavelengths sensor.

Similarly, the AFE44 × 0SpO 2 EVM from Texas Instruments is another kit intended for evaluating AFE4400 and AFE4490 devices. The AFE4490 is a complete analog front-end solution targeted for pulse-oximeter applications. The device consists of a low-noise receiver channel, a LED transmit section, and diagnostics for sensor and LED fault detection. The software of the AFE44 × 0SpO 2 EVM includes a GUI that allows the configuration of the I-V amplifier, the ambient light cancellation DAC, the analog filtering, the Analog-to-Digital Converter (ADC) sampling and LED pulse & intensity, among many other parameters. This is a very complete evaluation system with highly configurable capabilities that unfortunately only allows two wavelengths.

To the authors' knowledge, there are no platforms that allow the exploration of multiple signals, higher frequency and higher resolution with respect to the quality of the PPG acquisition.

3 Platform Design

The proposed platform is designed to tackle the limits of current devices for PPG analysis. It extends current platforms by proposing multiple wavelengths (more than 3), higher frequency, higher resolution while maintaining a 'nomadic' device.

(a) Transmittance pattern before encapsulation.

(b) Reflectance pattern encapsulated in a protective resin.

Fig. 3. The two first sensors manufactured for this study.

The device should capture and store raw data, as well as making embedded signal processing possible with a general purpose processor.

The platform is based on a PYNQ-Z1 board, incorporating a Xilinx Zynq SoC. The integrated ARM processor, equipped with an operating system, serves as an ideal sandbox for signal processing algorithm research. Leveraging the FPGA within the Zynq platform facilitates the migration of the acquisition process to dedicated hardware. This ensures the reliability of the acquisition while avoiding any adverse impact on the ARM processor. A custom electronic board complement the PYNQ-Z1 to control the physical sensor.

The platform uses an original physical sensor presented in Sect. 3.1, custom electronic design for LED power and signal acquisition explained in Sect. 3.2 and finally an integrated logic design as coordinator detailed in Sect. 3.3.

3.1 Physical Sensor Design

One of the primary goals of the advanced PPG sensor is to provide measurements with more wavelengths than the typical green/red/IR. Consequently, a new sensor has been designed with 8 LEDs allowing 8 wavelengths to be studied. It is based on non-encapsulated LEDs (300×300 μm), with the objective to minimize the sensor size. The sensor is covered with a protective resin.

With the constraints of size, market availability and price, the following 8 LEDs wavelengths have been chosen. A red LED operating at 660 nm and an IR LED operating at 940 nm allow the comparison with standard oximeters. LEDs operating at 542 nm, 593 nm, 620 nm, 762 nm, 807 nm, 888 nm are selected to estimate Carboxyhemoglobin and Methemoglobin.

Three sensor patterns consisting of 8 LEDs covering wavelengths from 542 to 940 nm were manufactured. The first sensor pattern on Fig. 3a was designed for use in transmittance, so contained 8 LEDs, with the photodiode on another PCB. The second sensor pattern on Fig. 3b was for use in reflectance, so the 8 LEDs surrounded a photodiode. The third sensor pattern is also used for reflectance, but the distance between the LEDs and the photodiode is greater, in order to be able to detect light that has penetrated deeper into the tissue. The reflectance

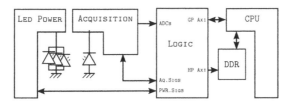

Fig. 4. Global schematic of the system.

mode allows the platform to be used in a wider variety of body sites compared to the transmittance mode.

3.2 Electronic Design

To enable sample measurements using these sensors, two specific electronic circuits have been designed. The primary goal of the first board is to provide stable and accurate currents to the LEDs. The second board performs fine measurements of the received current in the photodiode. These two circuits are connected to the LEDs, photodiode and logic as shown in the Fig. 4.

LED Power. The first electronic circuit drives the current in a specific LED using 2 signals. The first signal is the *enable* signal. This signal is composed of one wire per LED and is set to a logic 1 (here 3.3V) which causes the selected LED to light up. The second signal is the *pwm* signal. This signal is a Pulse Width Modulation (PWM) signal that control the amount of current to drive into the LED. The Fig. 5 represents the circuit design for a specific LED.

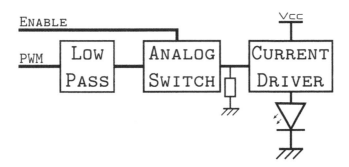

Fig. 5. Electronic LED power stage for one LED.

In this circuit, the *pwm* signal passes inside a low pass filter (an RC design) that transform the PWM into an analog signal. The analog switch injects this signal in the current driver based on the level of the enable signal. The analog switch chosen is a HCT4066-based circuit. A pull down resistor sets the output of the switch to ground when the switch is disconnected. Finally, the current driver section operates the current regulation. The schematic is based on an Operational Amplifier (OpAmp) and a transistor (TLC271CP and BC847CW). This circuit is duplicated 8 times in the complete circuit to drive the 8 LEDs independently.

Acquisition. The light emitted by the LED is then attenuated or reflected by the tissue of the subject and captured by the photodiode. Figure 6 shows the schematic of the capture circuit.

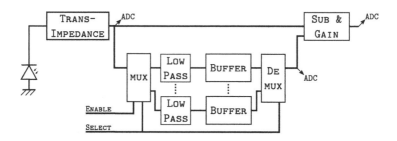

Fig. 6. Photodiode stage for data acquisition.

This schematic has 2 types of input signals and 3 ADC outputs. The ADC engines of the targeted architecture can select one of these 3 signals and selectively measure the three inputs. The 2 inputs are a 3-bit *select* that specify which LED is currently measured and a 1-bit *enable* signal that specify if the device is in calibration mode.

The first block is a transimpedance block composed of AD8021AR OpAmp that converts the current delivered by the photodiode to a voltage. This first voltage is measured (after a level adjuster) to one of the ADC pins of the Zynq chip. This voltage can be used to evaluate the ambient noise during a capture.

To maximize the signal precision, the signal is amplified. However, since the pulsating level of the signal is significantly lower than the steady component, the choice has been made to subtract a fixed offset value from the signal. This fixed offset value is determined at the start of the capture by taking the average value of the signal for each LED.

This operation is performed through the first analog Mux (74HCT4051) that allows the input signal to charge a low pass RC filter when the *calibration* signal is set. The signal is then buffered using an TLC271 OpAmp used as voltage follower, thus considerably increasing the discharge time of the filter capacitor. After calibration, DC-offset can be measured using the second analog Mux that outputs to an ADC.

Finally, the full signal and the offset are subtracted then amplified using an TLC271 OpAmp in differential amplification mode. Resistors are chosen accordingly to obtain the following output voltage:

$$V_{out} = gain * (V_{avg} - V_{in}) + 0.5 \tag{1}$$

Choosing an offset of 0.5V allows maximum signal variation, as the ADC has a range of 0–1 V.

3.3 Logic

The analog/digital interface is controlled by the FPGA, to operate the sensor's electronic and the ADC. The signal capture is not processor dependent and the FPGA guarantees synchronicity of the capture operation as well as less processing on the processor. The logic system is described in Fig. 7.

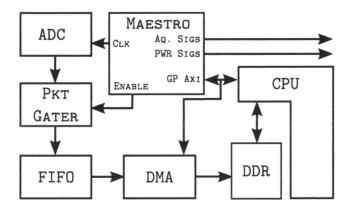

Fig. 7. Schematic of the FPGA design.

A global state machine called Maestro handles the capture process. It outputs the signals for the LED and photodiode electronics and pilots the capture of the ADC. The LEDs are switched on and off sequentially, with a negligible delay between them, so that the signals from the 8 wavelengths can be measured independently.

The ADC has a 12-bit resolution, but it outputs a 16-bit result. The least significant 4 bits in the result are not meaningful and can be used to reduce the quantization noise by averaging. The 16 bits result is transmitted to the user.

Data Management. The processor communicates with the Maestro through a general purpose AXI interface, by writing in specific registers. Samples are retrieved directly through the Double Data Rate? (DDR).

When performing the capture, the ADC outputs its captured data through an AXI Stream interface. This stream is then gated to allow the sample to flow only when a capture is performed. This is necessary to both prevent the ADC to output a sample at startup when it is not configured correctly, but also for debugging (performing dry runs).

When the capture starts, samples are buffered in a First in, first out (FIFO) queue before being copied to the DDR using a Direct Memory Access (DMA). When triggered by the processor, the DMA writes packets of samples to a given address in the DDR. High performance AXI ports of the Zynq architecture are used to achieve a high throughput.

Highly Configurable. The use of an FPGA also provides a wide range of sampling frequencies. The platform support sampling frequencies ranging from 200 Hz to 10 kHz, each measurement comprising a sample for the selected LEDs (up to 8). In addition, various acquisition parameters can be configured.

As mentioned before, the current in the LEDs can be specified separately for each LED. A custom software on the processor performs an exhaustive search for the best power value in the LED for each LED, in order to optimize dynamic range during acquisition. This can be used to adapt the platform to the experiment environment but also to the experiment subject.

The delay between turning on the LED and the measurement can be changed. There is a compromise to find between the settling time and fastness. Similarly, it is possible to configure the fixed delay between turning off the previous LED and turning on the next one. The numbers of LEDs used for an experiment can be chosen. Any combination that uses between 1 and 8 LEDs is allowed. The number of packets used for the calibration of the offset and the number of samples per packet can also be configured.

4   Evaluation

Several experiments were performed in order to evaluate the proposed platform. This section presents 2 of them, with an evaluation of the acquired signal quality and a first biomarker extraction.

The first experiment illustrates the added value of the AC extraction, i.e. the removal of the signal offset before acquisition. The advanced photodetector was used to capture a signal for 4 s with a sampling rate of 2000 Hz. The complete signal and the AC signal, both unfiltered, were normalized between 0 and 1 to compare them. Figure 8 presents the result for the signals, with focus on the 940 nm LED. The AC (removing the DC before acquisition) is less noisy than the full signal (removing the DC after the acquisition). The average variance on windows of 40 samples is *0.066* versus *0.476*.

The sum of the variances of the signals on windows with a size of 40 samples was calculated. The results are given in the Table 1 where V_n is the variance of the digital zoom and V_e is the variance of the electronic zoom and *lambda* the wavelength in nm.

For the second experiment, the subject was asked to repeat breathing cycles consisting of 3 s of inspiration followed by 3 s of expiration. We can observe the acquired AC signal on Fig. 9. The envelope of the signal has a periodicity

Table 1. Comparison of variance between complete and AC normalized signals.

λ	542	593	620	660	762	807	888	940
V_n	0.73	0.31	0.29	0.46	0.63	0.21	0.71	0.47
V_e	0.11	0.04	0.04	0.07	0.06	0.03	0.13	0.05

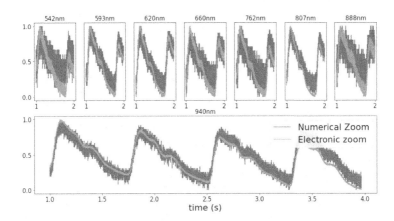

Fig. 8. Comparison of precision between complete and AC normalized signals.

of about 6 s, which corresponds to the experimental conditions. The breathing is thus easily extracted from the signal obtained with the sensor. The same acquisition without offset removal would be much more difficult to interpret.

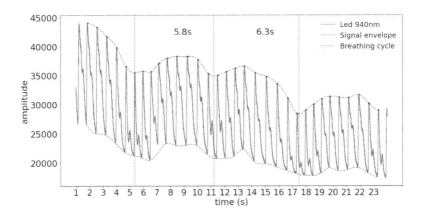

Fig. 9. Breathing cycle measurement using PPG.

5   Conclusion

In conclusion, this paper has centered on the development of an extensively configurable PPG sensor with accessible raw output data. This feature enhances flexibility in scientific investigations within the PPG domain. The direct access to raw data facilitates seamless comparison of research outcomes among scholars. Notably, our sensor empowers researchers to personalize critical parameters, such as the sample rate and the utilisation of multiple LEDs (up to 8). This flexibility opens avenues for exploring novel methods of calculating biomarkers.

Moreover, the adaptability of this sensor extends the possibilities for devising energy-efficient hardware strategies. By adjusting parameters based on the specific requirements and reducing energy consumption during periods of low information, novel approaches to energy-aware hardware design can be explored. Furthermore, the portability of this platform, coupled with its compatibility with a diverse array of body sites, makes it a versatile tool for both educational and research purposes.

Future work will explore the use of this new sensor and platform in a clinical trial, associated with a battery of other sensors. This trial will help validate the performance of the proposed platform. It will also enable new research into biomarker extraction from the acquired PPG signals.

References

1. Bitalino faqs. https://bitalino.com/documentation/faqs. Accessed 11 Nov 2022
2. Chang, K.M., Chang, K.M.: Pulse rate derivation and its correlation with heart rate. J. Med. Biol. Eng. **29**, 132–137 (01 2009)
3. Chen, L., Reisner, A.T., Reifman, J.: Automated beat onset and peak detection algorithm for field-collected photoplethysmograms. In: Conference of the IEEE Engineering in Medicine and Biology Society, pp. 5689–5692 (09 2009)
4. Makivic, B., Nikić, M., Willis, M.: Heart rate variability (HRV) as a tool for diagnostic and monitoring performance in sport and physical activities. J. Exercise Physiol. Online **16**, 103–131 (01 2013)
5. Solé Morillo, Á., Lambert Cause, J., Baciu, V.E., da Silva, B., Garcia-Naranjo, J.C., Stiens, J.: PPG EduKit: an adjustable photoplethysmography evaluation system for educational activities. Sensors **22**(4), 1389 (2022)
6. Tamura, T., Maeda, Y., Sekine, M., Yoshida, M.: Wearable photoplethysmographic sensors-past and present. Electronics **3**(2), 282–302 (2014). https://doi.org/10.3390/electronics3020282
7. Temko, A.: Accurate heart rate monitoring during physical exercises using PPG. IEEE Trans. Biomed. Eng. **64**(9), 2016–2024 (2017)
8. Vadrevu, S., Manikandan, M.S.: A robust pulse onset and peak detection method for automated PPG signal analysis system. IEEE Trans. Instrument. Measure. **68**(3), 807–817 (03 2019). https://doi.org/10.1109/TIM.2018.2857878
9. Zijlstra, W., Buursma, A.: Spectrophotometry of hemoglobin: absorption spectra of bovine oxyhemoglobin, deoxyhemoglobin, carboxyhemoglobin, and methemoglobin. Comp. Biochem. Physiol. B: Biochem. Mol. Biol. **118**(4), 743–749 (1997). https://doi.org/10.1016/S0305-0491(97)00230-7

sEMG-Based Gesture Recognition with Spiking Neural Networks on Low-Power FPGA

Matteo Antonio Scrugli[(✉)] [ID], Gianluca Leone[ID], Paola Busia[ID], and Paolo Meloni[ID]

University of Cagliari, Cagliari, Italy
{matteo.scrugli,gianluca.leone94,paola.busia,paolo.meloni}@unica.it

Abstract. Classification of surface electromyographic (sEMG) signals for the precise identification of hand gestures is a crucial area in the advancement of complex prosthetic devices and human-machine interfaces. This study presents a real-time sEMG classification system, exploiting a Spiking Neural Network (SNN) to distinguish among twelve distinct hand gestures. The system is implemented on a Lattice iCE40-UltraPlus FPGA, explicitly designed for low-power applications. Evaluation on the NinaPro DB5 dataset confirms an accuracy of 85.6%, demonstrating the model's effectiveness. The power consumption for this architecture is approximately 1.7 mW, leveraging the inherent energy efficiency of SNNs for low-power classification.

Keywords: Spiking Neural Networks · Real-time monitoring · Healthcare

1 Introduction

The continuous development of artificial intelligence and neural networks offers numerous opportunities for long-term monitoring and personalized applications in the healthcare domain. One area of considerable interest is the precise interpretation of surface electromyographic signals (sEMG) for accurate identification of hand gestures, a cornerstone for the evolution of sophisticated prosthetic devices and human-machine interfaces. Real-time monitoring of sEMG poses great challenges due to stringent accuracy and resource limitations in wearable/embedded systems, making it a key research area. Spiking Neural Networks (SNNs) emerge as a promising solution with energy-efficient, event-driven processing. However, exploiting the benefits of event-driven processing often requires specialized computational architectures.

Field Programmable Gate Arrays (FPGAs), due to their highly customizable hardware design, present themselves as suitable candidates for these computational tasks. Their design allows for the exploitation of sparse neuron firing patterns. The core Digital Signal Processor (DSP) slices are engineered to

This work was supported by Key Digital Technologies Joint Undertaking (KDT JU) in "EdgeAI Edge AI Technologies for Optimised Performance Embedded Processing" project, grant agreement No 101097300.

T. Dias and P. Busia (Eds.): DASIP 2024, LNCS 14622, pp. 15–26, 2024.
https://doi.org/10.1007/978-3-031-62874-0_2

Table 1. A comparative summary of relevant works in sEMG classification using Spiking Neural Networks.

Work	Dataset	Encoding	Classess	Accuracy	Device	Energy	Power	Mops
[3]	custom 4 subjects	Delta	4	84.8%	SpiNNaker	N.R[1]	1–4 W	N.R.
[4]	custom 8 subjects	Population	8	97.4%	N.R.	N.R.	N.R.	0.013
[10]	custom 5 subjects	HD-sEMG Decomposition	10	95%	Jetson	0.97 mJ	100 mW	7.97[*]
[12]	custom 10 subjects	Event-Differential	6	98.78%	N.R.	N.R.	N.R.	6.57
[11]	NinaPro DB5	Delta	12	74%	Loihi	246 mJ	41 mW	11.56[*]
Our	**NinaPro DB5**	**Delta**	**12**	**85.6%**	**FPGA**	**35.68 µJ**	**1.7 mW**	**2.336**

[1] Not Reported.
[*] Estimated from the paper.

adeptly handle a suite of arithmetic operations including addition, multiplication, and multiply-and-accumulate. On another front, BRAM (Block Random Access Memory) units, due to their adaptable design and size, are especially suitable for integrating SNN models and facilitating data access and management.

SNN-based systems, although an emerging technology, show promising potential in the classification of sEMG signals for hand gesture recognition, a key step toward the advancement of complex prosthetic devices and human-machine interfaces. This paper presents a real-time sEMG classification system for distinguishing hand gestures and investigates the integration of Spiking neural networks on the Lattice iCE40-UltraPlus FPGA, a device optimized for low-power applications, addressing the challenges and potentials in this area.

The main contributions can be summarized as three most relevant points:

- we present an SNN-based real-time classification system capable of discerning twelve distinct hand gestures using sEMG signals, achieving an accuracy of 85.6% that underscores the effectiveness of the model;
- we investigate the most efficient coding method to transmit the continuous sEMG signal in spike traces, evaluating the benefits of using delta modulation on the original signal, as well as on the first- and second-order derivatives, which enhances classification performance;
- we illustrate an efficient implementation on ultra-low-power Lattice FPGAs, consuming about 1 mW, which exploits the inherent energy efficiency of SNNs for low-power classification.

2   Related Work

Recent works from the literature demonstrate the efficiency of SNNs and their suitability for the gesture recognition problem based on the sEMG signal processing. In Table 1, we summarize the most relevant contributions, reporting the reference dataset, the data encoding scheme, the number of different gestures considered, and the classification accuracy. Moreover, we include the on-hardware efficiency of the proposed models, reporting the targeted platform and the measured energy and power consumption for inference execution.

The work of [3] demonstrates the use of the NeuCube spiking model for the recognition of 4 different hand gestures, reaching 84.8% classification accuracy. The reference data was acquired from 4 volunteering subjects, performing two-digit grasp, three-digit grasp, fist, and hand rest. The train of spikes provided as input to the SNN model was obtained through temporal difference encoding. Similarly, the work of [12] presents an SNN model reaching 98.78% accuracy in the recognition of 6 different gestures, based on the custom recordings of 10 subjects, whereas in [10] 95% classification accuracy was obtained in the classification of 10 hand gestures. The authors of [4] propose a different encoding scheme, exploiting a set of Gaussian filters to produce the train of spikes. They report 97.4% accuracy in the recognition of 8 different gestures based on the sEMG recorded from 8 subjects. As can be noticed, the listed works reference custom datasets, comprising different sets of hand gestures, therefore the comparison of the reported results in terms of achieved performance is complicated by the lack of a common data reference [3,4,10,12].

Due to this reason, we considered as our main reference the work of [11], which targeted the open-source NinaPro database, considering a subset of the data collected in the DB5, and particularly 12 gestures comprising flexion and extension of each finger, thumb adduction, and abduction, plus a rest class. The authors perform a design exploration for the selection of the optimal SNN model, comparing topologies including convolutional and fully connected layers. The finally selected model exploits the CUBA leaky-integrate and fire neuron and embeds two fully connected layers, reaching 74% classification accuracy. With this work, we aim at improving the classification performance achieved in [11], proposing a lean topology with a reduced number of required operations and relying on a novel version of the delta encoding scheme, where we include the information provided by the first and second derivatives of the signal.

Moreover, considering the applications of sEMG processing, which include prosthesis control, evaluating the efficiency of the proposed classifier is especially relevant. The SNN model presented in [3] was deployed on the SpiNNaker platform [6], providing real-time classification with power consumption ranging from 1 to 4 W. The model presented in [11] was evaluated on the Loihi [5], demonstrating 5.7 ms inference time, with 246 mJ energy per inference and 41 mW average power consumption for parallel execution on 12 cores. On the other hand, the model presented in [10] was deployed on the NVIDIA Jetson Nano platform, resulting in 9.7 ms inference time, with 0.97 mJ per inference and 100 mW average power consumption. Finally, direct measurements on target hardware are not provided for the models in [12] and [4], we thus report their complexity in terms of the number of required accumulate operations. To the best of our knowledge, the implementation presented for our proposed gesture recognition model results in the lowest power consumption among the referenced works, thanks to the specialized re-configurable processing architecture designed on the low-power Lattice FPGA and a limited network complexity in terms of the number of required operations.

3   Method

In this section, we delineate the specifics of our proposed system, showcasing the reference dataset, the encoding methodology for spike generation, the chosen SNN architecture, and the targeted FPGA-based processor adapted to sEMG signal processing.

3.1   Dataset

The data used in this study were derived from the NinaPro DB5 dataset [8], which includes sEMG and kinematic data collected from 10 subjects without amputations or disabilities performing 52 distinct hand movements alongside a resting position. The 52 gestures are divided into subgroups, E1, E2, and E3, in alignment with the work of [11], which provides the main reference for comparison, we selected the E1 subgroup comprising of 12 exercises. The labels are structured to capture different movements for each finger. For the index, middle, ring, and little fingers, there are two labels each, one for flexion and another for extension, making eight labels. The thumb adds four more labels for adduction, abduction, flexion, and extension.

Acquisition was performed using two Thalmic Myo Armbands and the sEMG signals were sampled at a frequency of 200 Hz. Each subject performed six repetitions of the specified movements, as demonstrated by movies on a laptop computer screen, with each repetition lasting approximately five seconds followed by a three-second rest interval.

For our analysis, we focused specifically on the sEMG data encapsulated in the "emg" variable of the dataset, which consists of 16 columns representing signals from 16 channels. The first 8 columns correspond to electrodes evenly distributed around the forearm at the height of the radio-humeral joint, while columns 9–16 represent the second Myo, with a 22.5-degree clockwise slope.

The labels are derived from the "restimulus" variable. While, as will be further explained in Sect. 3.3, the dataset was divided into training, validation, and testing subsets based on the "rerepetition" variable. The names "restimulus" and "rerepetitin" for the variables derive from the original dataset description.

3.2   Encoding

In the context of spiking neural networks, an important part of the process is to transform continuous sEMG signals into spike sequences. In this study, we use Delta modulation, a method commonly used in signal processing and telecommunications to transform analog signals into digital form.

Our approach is different from the usual methods found in the literature because we apply delta modulation not only to the raw signal but also to its first and second derivatives. Using the first derivative, we emphasize information related to the speed of muscle activations. Meanwhile, the second derivative helps us analyze the acceleration or deceleration of these activations, adding more depth to our analysis.

The mechanism of delta modulation generates two distinct traces, indicating either a positive or negative deviation of the signal from a designated threshold, δ, which in our setup, is assigned a value of 15. The signal traces obtained through delta modulation are used directly without any normalization. Through a series of tests, the δ value chosen was validated to effectively represent variations in the sEMG signal. The encoding method is described as pseudo-code in Algorithm 1.

Algorithm 1: sEMG Encoding with Delta Modulation

   Input: *sEMG signal*
   Result: *Two signals of spikes per channel, POS and NEG.*
1 *delta sample = First sEMG sample*;
2 *delta value*;
3 **while** *sEMG signal* **do**
4   **for** *sEMG channel* **do**
5   *sEMG sample*;
6   **if** *sEMG sample > delta sample + delta value* **then**
7   *delta sample = sEMG sample*;
8   *POS_channel.append(spike)*;
9   **else**
10   *POS_channel.append(no-spike)*;
11   **if** *sEMG sample < delta sample - delta value* **then**
12   *delta sample = sEMG sample*;
13   *NEG_channel.append(spike)*;
14   **else**
15   *NEG_channel.append(no-spike)*;

3.3 Segmentation

The segmentation process is a crucial step in preparing sEMG data for subsequent analysis and classification activities. This step involves sectioning continuous sEMG signals into segments or windows, which are then used for feature extraction and classification.

Initially, the repetitions within the dataset are allocated to training, validation, and testing subsets. Specifically, repetitions 1, 2, 4, and 6 are earmarked for training, repetition 3 for validation, and repetition 5 for testing.

The windowing process, a central aspect of segmentation, is governed by a set of parameters. Primarily, the window size and shift, which are set at 0.5 s and 0.1 s respectively, dictate the extent and overlap of the segments. Given the data frequency of 200 Hz, these time-based parameters are translated into sample-based metrics, yielding a window size of 100 samples and a window shift of 20 samples.

In the NinaPro DB5, each sample is originally labeled as either *exercise* or *rest*. To assign a single label to a window of samples, we established specific criteria based on the proportion of labels within each window. A window is classified as *exercise* if at least 80% of its samples are labeled as such. Conversely, a window is labeled as *rest* if 100% of its samples carry the *rest* label. Windows that do not conform to these criteria are excluded from the training phase.

To eliminate some noise present at the beginning of the signal, we imposed an initial delay of 2 s before the start of the windowing process, equivalent to 400 samples based on the data sampling rate.

3.4   SNN Topology and Training

We used the SLAYER [9] libraries to manage the neural network; as a matter of practical implementation, the selected neuron type is the LIF (Leaky Integrate and Fire).

Outlined in Eqs. 1 and 2, each neuron receives a sequence of spikes s_{in} as input, which, when subject to synaptic weight w multiplication, shape the membrane voltage v within the neuron. The voltage v gradually decreases, influenced by the decay factor α.

The SLAYER toolbox is essential for defining custom spiking neuron and synaptic behaviors. It was used in conjunction with the LAVA framework [1,2], which enhances GPU-based efficiency for SNN training and simulation.

$$\tilde{v}(t) = \alpha \cdot v(t) + \sum w \cdot s_{in}(t) \tag{1}$$

$$v(t+1) = \begin{cases} \tilde{v}(t), & \text{if } \tilde{v}(t) < \theta \\ 0, & \text{otherwise} \end{cases} \tag{2}$$

As shown in Eq. 3, the neuron emits an output spike $s_{out}(t)$ when its voltage exceeds a certain threshold θ. The output spike can be represented as:

$$s_{out}(t+1) = \begin{cases} 1, & \text{if } \tilde{v}(t) \geq \theta \\ 0, & \text{otherwise} \end{cases} \tag{3}$$

We carried out the training on Google Colab, leveraging the computational power of T4 GPUs. The Adam optimizer was chosen with a learning rate of 0.001 and batch sizes fixed at 32. The loss was assessed through the *slayer.loss* class, with a focus on *SpikeRate*-based calculations. We set a true rate target at 0.2 and a false rate at 0.03. To avoid overfitting, an early stopping technique was used. In our case, training is stopped if there is no improvement within 10 consecutive epochs.

The architecture of our network is assembled through a series of Dense layers, delineated in Table 2. Finally, through the support of the SLAYER [9] framework, we introduced axonal delays to simulate the propagation time of spikes along the axon, setting a maximum delay value of 62 time-steps.

Table 2. Topology and parameters of the proposed SNN.

Layer	Synapse	Neuron	Delay
Dense Layer 1	96	64	True
Dense Layer 2	64	128	True
Dense Layer 3	128	64	True
Dense Layer 4	64	13	False

3.5   System Overview

Figure 1 illustrates the schematic representation of the system implemented. The core control tasks are orchestrated by a RISC-V processor.

The input signal acquired through the SPI interface is first processed through the *encoding* stage, where it is translated into spike trains using delta modulation, as detailed in Sect. 3.2. At this point, the dataflow progresses to the *segmentation* block, which defines a partitioning of the data into windows, as described in Sect. 3.3.

Thereafter, the segmented spike trains are channeled into the SNN processor. The processor analyzes the spike train inputs to extract meaningful information about the gestures represented by the sEMG signals. After each inference, a *voting* phase is initiated to examine the neural network outputs and determine the final classification.

The classification output is then propagated through the UART interface, concluding the structured data processing path from acquisition to classification.

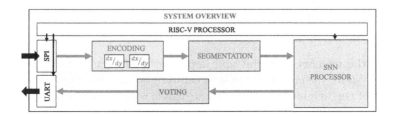

Fig. 1. Overview of the sEMG monitoring system.

Figure 2 outlines the architecture of the SNN processor utilized in our system, drawing inspiration from the approach delineated in [7]. The input data is represented by the spike trains channeled from the encoding module. The processor hosts two *neuron* modules, each capable of accumulating four 8-bit synaptic weights per cycle, aiding in the computation between synapses and weights.

Both neuron modules are interfaced with a weight memory and a spike memory. The weight memory, populated during initialization with the network weights derived from training, and the spike memory, which stores the results of previous inferences for all neurons in the SNN, provide the data needed for the calculation. An *address generator* dynamically computes the memory addresses, assisting each of the two available modules in processing a distinct layer of the SNN.

A *spike stack* works with the address generator to point out where the active spikes are in the memory, helping to skip over the inactive synapses. The spikes that come out of this process are sent to the voting module to be counted, determining the final classification.

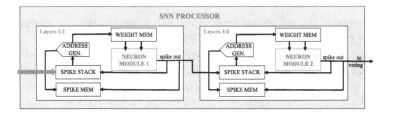

Fig. 2. Overview of the sEMG SNN processor.

4   Experimental Results

This section delineates the findings from our experimental analysis, evaluating both the classification accuracy and the on-hardware inference efficiency of the proposed sEMG monitoring system.

4.1   Classification Evaluation

To reduce the memory footprint and improve the efficiency of the designated hardware system, the chosen model was subjected to quantization, reducing the resolution of the weights to 8 bits with negligible accuracy drop. This approach resulted in a classification accuracy of 85.6% on the test set, obtained with an output post-processing stage, which allows us to filter out single-spot classification, through a voting mechanism that discards classifications resulting in non-consecutive labels, by restoring the last valid output classification.

Table 3 represents the confusion matrix, providing a detailed analysis of classification performance across classes. As can be noticed, a relevant portion of the errors, around 45%, is located in the first column and the first row of the matrix, thus involving the distinction between the rest class and all the remaining gestures.

In this regard, we also evaluated the impact of the classification errors registered during the start of each gesture, knowing that the first windows include a certain portion of *rest* samples and *gesture* samples. In detail, we introduced a tolerance of 200 ms around the gesture onset, where we considered as acceptable classifications both the *rest* and the specific *gesture* class. Considering this additional tolerance, the overall classification accuracy achieved is 87.01%.

A graphical representation of an example of classification output is provided in Fig. 3. The plot reports two repetitions of the ring flexion and extension gestures, separated by a resting phase. The classification output is indicated on the plot, and placed according to the predicted class. As outlined in Sect. 3.3, dataset repetitions are assigned to training (repetitions 1, 2, 4, 6), validation (repetition 3), and testing subsets (repetition 5).

Table 3. Confusion matrix reporting classification performance on the test set.

True Labels	Rest	Idx Flx	Idx Ext	Mid Flx	Mid Ext	Ring Flx	Ring Ext	Lit Flx	Lit Ext	Thm Add	Thm Abd	Thm Flx	Thm Ext
Rest	7078	36	1	28	43	1	4	26	1	39	86	2	32
Idx Flx	62	268	30	15	21	5	17	11	0	0	2	2	2
Idx Ext	31	23	224	8	12	3	1	0	0	0	2	0	0
Mid Flx	30	31	3	285	14	7	8	13	3	3	0	0	0
Mid Ext	17	4	3	0	244	8	4	5	7	0	2	0	0
Ring Flx	43	8	2	12	12	258	2	14	2	3	0	2	2
Ring Ext	39	30	3	17	36	8	218	7	3	2	0	0	10
Lit Flx	48	5	5	6	13	18	26	237	7	10	6	3	11
Lit Ext	52	0	0	5	5	10	13	13	253	0	13	7	0
Thm Add	7	10	0	0	0	0	10	4	0	161	6	70	7
Thm Abd	71	2	4	0	0	0	2	3	3	7	190	17	19
Thm Flx	21	0	4	0	0	0	0	0	0	64	2	166	10
Thm Ext	32	0	10	2	2	0	0	7	2	9	3	6	252

Predicted Labels

4.2 Sparsity

SNNs naturally exhibit sparsity, since in a specific instance only a small subset of neurons are engaged in firing or communication. Sparsity is defined as the ratio of inactive spikes to the maximum possible spike count within the network, reflecting the network's level of inactivity or quietness. In our context, taking the test set into consideration, a sparsity value of 90.99% was calculated, indicative of a mere 9% average presence of potential active spikes. The platform's processing elements allow inference to be calculated for four peaks in parallel. As a consequence, spikes are processed in quartets, therefore it becomes critical to assess the extent to which the inherent sparsity of the model architecture can be exploited on the designated hardware and translated into performance efficiency. As a result of such grouping of spikes, it is necessary to introduce a

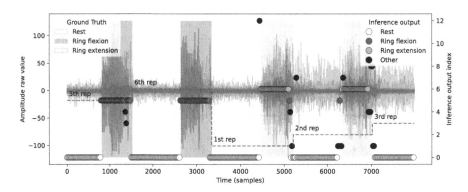

Fig. 3. Overview of the EMG signal classification for different repetitions of the rest, ring flexion, and extension gestures. The ground truth is highlighted in the background, whereas the classification output is reported as spots overlapped to the signal plot. Classification errors are reported in black.

new element, called *sparsity-hw*, which is defined by the Eq. 4.

$$Sparsity\text{-}hw = \left(1 - \frac{\sum_{i=1}^{L}\sum_{j=1}^{G_i} g_{ij}}{\sum_{i=1}^{L} G_i}\right) \times 100 \tag{4}$$

where:

- L is the total number of layers in the network,
- G_i is the number of spike groups in the i-th layer,
- g_{ij} indicates whether there's at least one active spike in the j-th group in the i-th layer (1 if at least one spike is active, 0 otherwise).

In our scenario, a *sparsity-hw* value of 77.11% was achieved.

4.3 Power Consumption

The system mainly switches between two operating states: the active inference phase and the idle phase. In the active inference phase, the FPGA is actively engaged in analyzing the spike data through the Spiking Neural Network (SNN), classifying the sEMG signals into the corresponding gesture categories. In the idle phase, on the other hand, the system goes into a low-power mode, greatly reducing power consumption. The average power consumption of the system can be assessed over the recurrent 100 ms inference interval. During the active phase, which includes inference, data acquisition, and encoding time and lasts for 3.921 ms, the power consumption is equal to 12.011 mW. In contrast, during the idle state, the power consumption drops to 1.288 mW. The average power over the 100 ms interval can be computed using Eq. 5:

$$P_{\text{average}} = \frac{(P_{\text{active}} \cdot T_{\text{active}}) + (P_{\text{idle}} \cdot T_{\text{idle}})}{T_{\text{total}}} = 1.708 \text{ mW} \tag{5}$$

To acquire power metrics, we used a Digilent Analog Discovery 2 oscilloscope in conjunction with three shunt resistors of 1.0 ± 0.01 Ω to measure power metrics. These resistors were installed across the three distinct power inputs on the Lattice iCE40UP5k FPGA, identified as Vcore, VCCIO0&1, and VCCIO2. The internal components of the FPGA receive a 1.2 V supply from Vcore, whereas the I/O pins are powered by 3.3 V from the VCCIO0&1 and VCCIO2 circuits.

4.4   Discussion

Our system, implemented on an FPGA, demonstrates a significant advantage in terms of power efficiency, consuming about 1.7 mW. This is substantially less than those mentioned in Sect. 2, such as [3] with a range of 1–4 W and [10] with 100 mW on Loihi.

Focusing on the NinaPro DB5 dataset, which is common between our work and [11], provides a fair ground for performance comparison. In terms of accuracy, our system achieves 85.6%, which is notably higher than the 74% reported by [11]. Nonetheless, the model proposed in this work requires the execution of 2.336 million of accumulate operations (MOPS) per classification, which is lower than the 11.56 MOPS estimated for the execution of the SNN proposed by [11].

In summary, our work stands out in terms of power efficiency and accuracy, representing a robust solution for real-time gesture recognition using sEMG data. While there is a trade-off in terms of OPS, the significant gains in accuracy and power efficiency underscore the effectiveness of our approach.

5   Conclusion

We introduced a real-time sEMG signal classification system aimed at the precise identification of hand gestures. By employing a Spiking Neural Network (SNN) for classification, the work taps into the inherent efficiency of event-based computation, providing a low-power solution apt for implementation on edge devices. The proposed model, when evaluated on the Ninapro DB5 dataset, exhibited a commendable classification accuracy of 85.6%, thereby substantiating the effectiveness of the SNN in recognizing twelve distinct hand gestures. The architecture, tailored for the Lattice iCE40-UltraPlus FPGA, displayed remarkable efficiency in terms of both computation and power consumption. The quantized version of the model, aimed at reducing the memory footprint and enhancing the operational efficiency on the hardware, maintained a high classification accuracy, showcasing the feasibility of deploying such models on resource-constrained platforms. Power examination demonstrated an average power consumption of 1.7 mW, underscoring the energy-efficient characteristic of the proposed system. This work, therefore, lays a solid foundation for the creation of energy-efficient and effective real-time sEMG-based gesture recognition systems, paving the way for advanced human-machine interfaces and prosthetic control.

References

1. Lava-DL SLAYER (2023). https://lava-nc.org/lava-lib-di/slayer/slayer.html. Accessed 8 Jan 2023
2. LAVA-NC (2023). https://lava-nc.org/index.html. Accessed Jan 8 2023
3. Behrenbeck, J., et al.: Classification and regression of spatio-temporal signals using NeuCube and its realization on spinnaker neuromorphic hardware. J. Neural Eng. **16**(2), 026014 (2019)
4. Cheng, L., Liu, Y., Hou, Z.G., Tan, M., Du, D., Fei, M.: A rapid spiking neural network approach with an application on hand gesture recognition. IEEE Trans. Cogn. Dev. Syst. **13**(1), 151–161 (2021). https://doi.org/10.1109/TCDS.2019.2918228
5. Davies, M., et al.: Advancing neuromorphic computing with Loihi: a survey of results and outlook. Proc. IEEE **109**(5), 911–934 (2021). https://doi.org/10.1109/JPROC.2021.3067593
6. Furber, S.B., et al.: Overview of the spinnaker system architecture. IEEE Trans. Comput. **62**(12), 2454–2467 (2013). https://doi.org/10.1109/TC.2012.142
7. Leone, G., Raffo, L., Meloni, P.: On-FPGA spiking neural networks for end-to-end neural decoding. IEEE Access **11**, 41387–41399 (2023). https://doi.org/10.1109/ACCESS.2023.3269598
8. Pizzolato, S., Tagliapietra, L., Cognolato, M., Reggiani, M., Müller, H., Atzori, M.: Comparison of six electromyography acquisition setups on hand movement classification tasks. PLoS One **12**(10), e0186132 (2017). https://doi.org/10.1371/journal.pone.0186132
9. Shrestha, S.B., Orchard, G.: Slayer: spike layer error reassignment in time. In: Proceedings of the 32nd International Conference on Neural Information Processing Systems, pp. 1419–1428. NIPS'18, Curran Associates Inc., Red Hook, NY, USA (2018)
10. Tanzarella, S., Iacono, M., Donati, E., Farina, D., Bartolozzi, C.: Neuromorphic decoding of spinal motor neuron behaviour during natural hand movements for a new generation of wearable neural interfaces. IEEE Trans. Neural Syst. Rehabili. Eng. **31**, 3035–3046 (2023). https://doi.org/10.1109/TNSRE.2023.3295658
11. Vitale, A., Donati, E., Germann, R., Magno, M.: Neuromorphic edge computing for biomedical applications: gesture classification using EMG signals. IEEE Sens. J. **22**(20), 19490–19499 (2022). https://doi.org/10.1109/JSEN.2022.3194678
12. Xu, M., Chen, X., Sun, A., Zhang, X., Chen, X.: A novel event-driven spiking convolutional neural network for electromyography pattern recognition. IEEE Trans. Biomed. Eng. **70**(9), 2604–2615 (2023). https://doi.org/10.1109/TBME.2023.3258606

Scalable FPGA Implementation of Dynamic Programming for Optimal Control of Hybrid Electrical Vehicles

Frans Skarman[✉][iD] and Oscar Gustafsson[iD]

Department of Electrical Engineering, Linköping University, Linköping, Sweden
{frans.skarman,oscar.gustafsson}@liu.se

Abstract. Dynamic programming (DP) can be used for optimal control of hybrid electric vehicles but requires a large number of computations to be performed. As many of these computations can be performed in parallel, FPGAs are an interesting platform for executing the dynamic programming algorithm. This paper presents a scalable architecture for performing dynamic programming on FPGAs using a pipelined model of a hybrid electric vehicle (HEV). The proposed architecture supports multiple parallel model execution units and is scalable to support a configurable number of units, inputs, states, and time steps. The run time of the optimization process is shown to be improved significantly compared to a CPU implementation. With four parallel model execution units, the design runs in about 1.5% of the time required for an Intel Xeon W-1250 CPU. This shows that DP-based optimal control is feasible for HEVs and that FPGAs can be used to achieve it.

1 Introduction

Dynamic programming can be used to improve the fuel efficiency of vehicles, both those with conventional engines [5] and Hybrid Electric Vehicles (HEVs) [11]. However, due to the computational complexity, Dynamic Programming (DP) is often only applied for off-line analysis and to generate benchmark performance results [18]. By executing the algorithms on Field Programmable Gate Arrays (FPGAs), the large amount of parallelism often available in the problem can be exploited to enable improved performance and potentially enable real-time use of the algorithms. The DP algorithm consists of a model of the system being optimized, and the environment which executes that model. The focus of this paper is the execution environment which must feed the model with inputs, and manage the results. For maximum performance, it is useful to allow several pipelined model evaluations to be performed in parallel which requires the control logic to provide inputs to, and store results from all those evaluations.

The main contribution in this paper is an FPGA architecture for performing DP for HEV optimal control. When paired with a pipelined model of the vehicle such as one generated by high-level synthesis, the system provides a significant speedup over a CPU, which enables real-time use of the optimization algorithm.

T. Dias and P. Busia (Eds.): DASIP 2024, LNCS 14622, pp. 27–39, 2024.
https://doi.org/10.1007/978-3-031-62874-0_3

The rest of the paper is structured as follows: in Sect. 2, the theoretical background of DP is given, specifically focusing on its use for optimal control and describing the algorithm to be implemented. Related works are discussed in Sect. 3. In Sect. 4, the proposed architecture is presented and the performance is discussed in Sect. 5. Synthesis results are presented in Sect. 6. Finally, some conclusions are drawn in Sect. 7.

2   Dynamic Programming

Dynamic Programming (DP) is an efficient exhaustive search method that solves optimization problems by recursively solving a set of simpler sub-problems [1]. When applied to optimal control, as described in [3], a discrete-time dynamic system is given by

$$x_{k+1} = f(x_k, u_k, w_k), \qquad c_{\text{input}} = g(x_k, u_k, w_k), \qquad k = 0, 1, \ldots, N-1, \quad (1)$$

where k is a discrete time index, x_k is the state of the system, u_k is an input to apply to the system, w_k is a disturbance or noise parameter. Finally, N is the search horizon, i.e., how far into the future to look ahead. The function f computes the resulting state when applying the inputs and disturbance in state k, whereas g computes the cost of applying these inputs.

In the HEV case, x_k is the current state of the vehicle: its velocity and battery state of charge, u_k is the control signal for the vehicle, for example: the desired combustion engine torque, electric motor torque, and gear. Finally, w_k is used to model external disturbance such as the wind, slope of the road or even traffic.

The goal of the optimization process is to find a sequence of inputs $u_{0..N}$ which minimize the cost of reaching the end of the journey. In this case, the cost is a configurable function of the fuel spent and the time taken to reach the destination. In practice, the optimization is done over a finite horizon N, rather than the whole journey. This is done because the conditions modeled by w_k will change over time, requiring re-evaluation of the optimal inputs. That is, even if an optimal control input sequence can be determined before the journey starts, e.g., the wind and traffic conditions are likely to change.

To compute the optimal input sequence, the inputs and states are discretized and the cost of reaching the goal from each state, $J[k][x_k]$, is computed. This array J is called the cost-to-go map. The cost of reaching a state can be computed recursively starting from the last time step where the cost of reaching the goal state is 0 and the cost of other states is infinite. At each time step k, the model is evaluated for all inputs u_k in order to compute which state x_{k+1} the inputs reach. The cost of reaching the goal from state with input u_k x_k is the sum of the cost of applying u_k and the cost of x_{k+1}. The cost of state x_k is then the minimum of these input-state cost pairs.

Note that the resulting state x_{k+1} of applying an input u_k is not guaranteed to be exactly at a discretization step. Therefore, for continuous states, the cost $J[k+1][x_{k+1}]$ must be approximated. Here, we use linear interpolation to determine the cost based on the adjacent discrete states. To simplify interpolation, it is assumed that the state space is bounded and discretized uniformly.

Algorithm 1 . Dynamic programming algorithm for computing the cost-to-go map J

$J[N][x_N] \leftarrow g_N(x_N) \quad \forall x_N \in S_N$
for all time steps $k \in (N-1)\dots 0$ **do**
  for all states x_k **do**
    $c_{\text{best}} \leftarrow \infty$
    for all inputs $u_k \in U_k(x_k)$ **do**
      $x_{k+1} \leftarrow f(x_k, u_k, w_k)$
      $c_{\text{input}} \leftarrow g(x_k, u_k, w_k)$
      $c_{\text{state}} \leftarrow \text{interpolate\_cost}(J[k+1], x_{k+1})$
      $c \leftarrow c_{\text{input}} + c_{\text{state}}$
      if $c < c_{\text{best}}$ **then**
        $c_{\text{best}} \leftarrow c$
      end if
    end for
    $J[k][x_k] \leftarrow c_{\text{best}}$
  end for
end for

2.1   The Dynamic Programming Algorithm

Algorithm 1 is used to compute the optimal cost-to-go-map J. For each state in each time step k, all inputs are evaluated by computing the resulting state x_{k+1} and the cost of the input c_{input}. Using that resulting state, the cost for the rest of the time steps is looked up in the cost-to-go map J based on the next time step, $k+1$.

In Algorithm 1, we can also see where parallelism is possible. In the outer loops, there are data dependencies on the cost-to-go array J, but because the middle loop only reads from $J[k+1]$ and writes to $J[k]$, each iteration of the state loop for one time step can be computed in parallel. There is a further opportunity for parallelism in the innermost loop which computes inputs. It has a data dependency on the c_{best} value, but because it computes a minimum value, the accumulation can be done in parallel.

In this work, the body of the state loop is fully pipelined which allows one iteration to complete each clock cycle with stalls between time steps. As is shown later, the duration of these stalls is very low and has minimal impact on the run time of the algorithm. In order to make further performance gains, the execution of the middle loop is split across a configurable number of pipelined execution units.

2.2   Next State and Input Cost

In this paper it is assumed that hardware for computing the next state, f, and input cost, g, is available. The hardware for evaluating the model can be implemented in multiple ways. In [14,15], the authors presented a high level synthesis tool tailored to this sort of vehicle model. However, In this work, we used Vitis HLS to generate the hardware for the model.

In order to make the hardware implementation efficient, the states x produced and consumed by f and g must be discretized by equidistant steps. It is also advantageous for the distance between discretization steps to be powers of two, and in problems where this is not the case already, it is up to the model to scale its inputs and outputs accordingly.

3   Related Work

Dynamic programming is a broad technique with many applications and implementations. However, due to the broad nature, an architecture suitable to solving one problem is not necessarily suitable for solving all of them. Several papers investigate the use of FPGAs to accelerate genome sequencing, for example [6,10,12,17]. All these techniques traverse the state space "diagonally" using systolic arrays, a technique which does not work for the control problem described here where dependencies between states are between time steps. In [9], an FPGA implementation of DP for *XML twig matching* is presented. Like the genome sequencing case, the structure of the problem is fundamentally different from the control application, which means the architecture is not useful here.

A problem more closely related to the control problem discussed here is that of non-linear Model Predictive Control (MPC). There, the goal is also to find an optimal sequence of inputs over a fixed horizon which minimizes some cost function. A common technique for solving this is *explicit MPC* where control laws for each state are computed off-line, making the on-line implementation a simple function evaluation, often expressed as a lookup table of linear gains [2]. Being a simple lookup table allows very efficient on-line computation, but it only works when the state space is not changing. This rules out the explicit *MPC* method for the problem discussed in this paper, where for example, the allowed states can vary as a function of time (e.g., speed limits). Explicit MPC on FPGA is a well-researched field, see, e.g., [7] and [4]. Finally, [8] is a survey of linear MPC on FPGA. However, while sharing a similar formulation, linear MPC relies on the underlying model being linear, which is not necessary in this work.

Zhu *et al.* present a Graphics Processing Unit (GPU) implementation of DP for control of HEV [19] which is very similar to the HEV optimization discussed here. This implementation is over 90% faster than a CPU implementation of the same problem. A direct comparison of run time of the GPU implementation and the FPGA implementation presented here is impossible however, as the exact details of the models used are not identical, a different state space is used and the GPU implementation takes traffic lights into account.

4   Proposed Architecture

Figure 1 shows an overview of the proposed hardware architecture for performing the cost-to-go calculation from Algorithm 1. The architecture consists of a

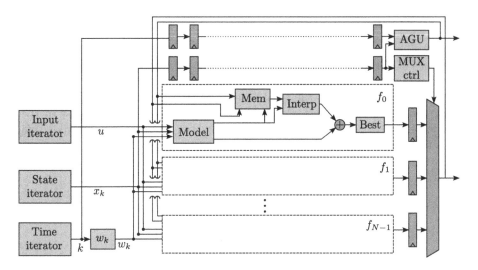

Fig. 1. Proposed architecture to perform dynamic programming using F parallel computation units.

number of computation units labeled f each containing the hardware required to execute the innermost loop body in algorithm.

The input generator block is implemented using cascaded counters which iterate through the time steps, states, and inputs to be evaluated. For parallelism, the state space is divided into M equally sized regions, and the input iterator iterates over one such region, with offsets being added to the state input for each functional unit. For example, if a 1D state space with ten discrete states is computed using two functional units, the input iterator iterates from $x = 0$ to 4 and the states passed to the functional units are x and $x + 5$, respectively. The input generator can also support configurable bounds on the state space.

The choice of state to parallelize can have an impact on the resource usage. For example, if the state space is split such that one functional unit always takes the same branch in an if-statement, the other branches can be optimized away from that unit. If word lengths are selected based on variable bounds, the word lengths can also sometimes be reduced with clever splitting of the state space. However, the effectiveness of these optimizations depends heavily on the details of the model and is out of scope for this work.

As will be explained in the next section, each functional unit has a local copy of the current and previous states. The current state is written back with the $J_{\text{writeback}}$ signal. The AGU block controls the writeback MUX and generates the addresses for the optimal cost memory, and the memories integrated into each functional unit. In order to maintain a high clock frequency, the memory and interpolation blocks in the model execution units are pipelined.

4.1　Functional Unit Structure

Figure 2 shows an overview of the layout of each functional unit. It consists of the model block which computes f and g to give the resulting state x_{k+1} and cost of applying the specified input c_{input}. The cost is added to the cost of the resulting state c_{next} to get the total cost to go of the input: c_{total}. The cost is finally compared to the best cost for the current state c_{best}. The cost of the resulting state c_{next} is, as mentioned previously computed through interpolation which is done using the memory- and interpolation-blocks.

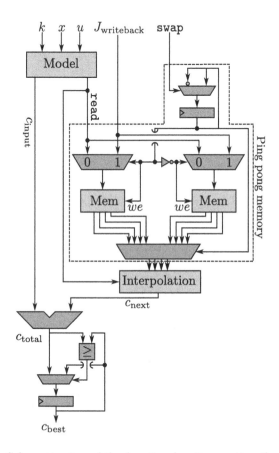

Fig. 2. Schematic view of the functional unit executing the model.

Each functional unit has two memories, one which stores the costs of the time steps which are currently being computed, and one which stores the costs of the previous state which is used to compute c_{next}. Between time steps, the role of the memory blocks are swapped.

Interpolation in 2D is done through bilinear interpolation which computes the value of a function f at a point x, y as

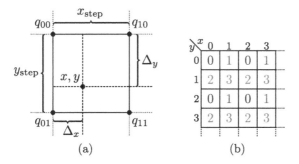

Fig. 3. (a) Values used for interpolation. (b) Distribution of cost-to-go values across four memories to allow reading four values simultaneously.

$$f(x,y) = q_{00} + \frac{\Delta_x \Delta_y}{x_{\text{step}} y_{\text{step}}}(q_{00} - q_{01} - q_{10} + q_{11}) + \frac{\Delta_x}{y_{\text{step}}}(q_{01} - q_{00}) + \frac{\Delta_y}{x_{\text{step}}}(q_{10} - q_{00}).$$
$$(2)$$

where Fig. 3a illustrates the involved variables. The q_{xy} values hold the value of the function at the four nearest points to x, y. The distance between the interpolated point and the known points is denoted by Δ_x, δ_x, Δ_y and δ_y, and the distances between the known points are denoted by x_{step} and y_{step} which, as equidistant discretization is used, are constant. When these are selected as powers of two, the division will be simplified to a bit-shift in fixed-point arithmetic.

The input state coordinates are divided in three, for x: X, X_{lsb} and Δ_x. The n most significant bits together make up X and X_{lsb} where X_{lsb} is a single bit, and the rest of the bits form Δ_X. In an equidistantly discretized state space with 2^n states, as is assumed here, X and X_{lsb} together with the corresponding y values form the index for q_{00}.

Computing (2) must be done every clock cycle, and each computation requires reading four values q_{xy} from the previous time step. This means that a memory with one or two ports is not sufficient. As the points are always in a square of neighboring points, one of the x indices will be even and one will be odd, a property which also holds for the y indices. This means that each memory in Fig. 2 can be composed by four individual memories, each containing a quarter of the total points. Figure 3b shows how the values are distributed across the four memories. For example, memory 0 contains the values of all points where both the x and y indices are even, and memory 2 contains the values of all points where x is even but y is odd.

The hardware used to index the memories in order to get q_{xy} is shown in Fig. 4. The inputs x and y are the indices where q_{00} would be stored if the module was implemented as a 2D array. The blocks Mem0 through Mem3 are memories corresponding to those shown in Fig. 3b. The variables x_o, x_e, y_o and y_e denote the indices to use for the odd and even memories respectively, i.e. the index for memory 0 is computed from x_e and y_e as it contains values where both x and

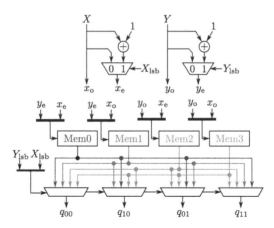

Fig. 4. Memory unit inside each functional unit with 2D state space. The boxes labeled with numbers are memories corresponding to the cells shown in Fig. 3b.

y are even. In order to get the final address for the memory, the corresponding indices are concatenated together as required for each memory. Finally, the four output values are assigned to the outputs depending on the odd and evenness of the x and y coordinates, as given by X_{lsb} and Y_{lsb}.

Single-port memories are sufficient as no reads occur when the memory block is written to. In the interest of clarity, the read ports have been omitted from Fig. 4. The same indexing hardware as is used when reading is used when writing values, and the write enable signal to each memory is computed based on the evenness of the write address.

5   Performance Characteristics

The execution time of the optimization process depends on the number of states S, the number of inputs U, the horizon N, the depth of the model pipeline D, the number of execution units P, and finally the latency of the interpolation block I. To compute the execution time, it is helpful to visualize the "critical path" of the computation as is shown in Fig. 5. In it the execution trace of a model with 6 states $s_{0..5}$, 3 inputs $u_{0..3}$ and 3 time steps is shown when executed on a model with a pipeline of depth 4, with two parallel execution units. Each tile corresponds to a stage in the pipeline being busy, and the highlighted tiles are the critical path determining the number of cycles required to compute the result. All states in a single time step can be computed in parallel, and in each state, each input must be evaluated.

In order to compute the total cost of a state, the costs of all the states in the previous time step must be available. Because the memories have a single write port, the cost of several states can not be written in parallel, so after the computation of the states, $I - 1$ cycles are used for computing the final costs and P cycles are used for writing those costs into the memory, shown in the w

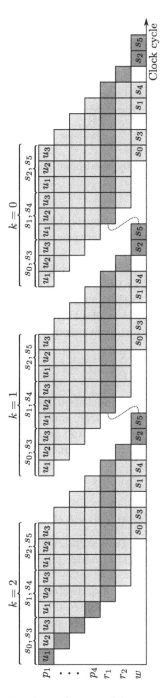

Fig. 5. Example of an execution trace for a model executed by two parallel execution units with four pipeline stages each (p_{1-4}) and two cycles of latency for reading and interpolating the previous cost (r_{1-2}). The model is simulated over three time steps k with six states s_{0-5} and three inputs u_{1-3} in each state. The highlighted tiles denote the critical path which limits the execution time.

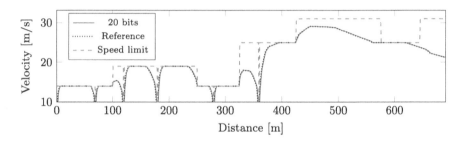

Fig. 6. Velocity profile computed by fixed-point model and reference floating-point model.

column of the figure. Once this writing is done, the interpolation of the next state cost can start. This means that between each time step, the pipelines must be stalled for P cycles.

In order to evaluate each input U in each state S across P execution units, $U \cdot \left\lceil \frac{S}{P} \right\rceil$ cycles are required. After that, $P + I - 1$ cycles are required for writing the results back. There are N such combinations to evaluate, and P cycles are required before the first output is available from the pipeline. This means that the total time taken to evaluate the optimization problem is

$$N \left(U \cdot \left\lceil \frac{S}{P} \right\rceil + P + I - 1 \right) + D. \tag{3}$$

As previously mentioned, the model used for this work has two states: the vehicle velocity and battery state of charge, and three inputs consisting of electric motor torque, combustion engine torque and gear. The states are both split into 30 discretization steps, as are the motor torques. The gear has six discrete steps. This means that

$$S = 30 \times 30 = 900$$

and

$$U = 30 \times 30 \times 6 = 5400.$$

The distance is discretized into 10 m steps, with a horizon of $N = 512$ steps resulting in a 5.12 km horizon. The model pipeline depth, D, is 894 and the latency of the interpolation block, I, is 10.

It is worth noting that the amount of discretization and length of the horizon is somewhat arbitrary. It provides useful results from the model, but the effect of modifying the number of discretization steps on the resulting total cost is left as future work.

6   Synthesis Results

The proposed architecture was implemented in the Spade hardware description language [13,16] and synthesized for the Xilinx Virtex UltraScale+ xcvu13p-fhga2104-3-e FPGA using Vivado 2022.1. The cost is stored as a 32-bit value with 20 fractional bits, and the x and y points are 32-bit values with 27 fractional bits [15]. Δ_x and Δ_y are each 29 bits wide with 27 fractional bits. In addition, the vehicle model described in [15] was synthezised using Vitis HLS 2022.01 using 32-bit fixed-point values with 20 fractional bits for the computations. As shown in Fig. 6, 20 fractional bits give similar performance to a reference floating-point model.

Table 1. Resource Usage on Virtex UltraScale+

EUs	CLB	DSP	BRAM	URAM	$f_{\max}$ [MHz]	CCs required	Run time [s]
1	10196	515	238	154	370	$2.488 \cdot 10^9$	6.73
2	21384	1030	476	154	336	$1.244 \cdot 10^9$	3.70
4	41654	2060	952	154	335	$6.220 \cdot 10^8$	1.85
6	62627	3090	1428	154	288	$4.147 \cdot 10^8$	1.44
9	94874	4635	2142	154	272	$2.764 \cdot 10^8$	1.02

Table 1 lists the resource usage, maximum clock frequency, number of clock cycles required for a full optimization according to (3), and the total runtime for the proposed architecture. It is clear that the LUT, DSP, and BRAM usage scales linearly with the number of execution units. The URAM is used for storing the final cost-to-go map, so it is unaffected by the number of execution units. In all cases, the maximum frequency is limited by the model. As the model hardware is the same, the reduction in frequency is likely caused by increased routing congestion.

The run time of the model is significantly improved over the CPU version which runs in 126 s in a single thread on an Intel Xeon W-1250 CPU. This means that the run time of the FPGA implementation with four execution units is less than 1.5% of the CPU implementation with even more performance improvement being possible with nine units, albeit at a much higher resource cost. For comparison, [19] accelerates a similar algorithm on a GPU. Their resulting design runs in 6% of the time for the corresponding CPU model. In addition to a significant speedup, run-times in the seconds range enable real-time use of the optimization algorithm, which has traditionally only been used for off-line analysis [18].

7   Conclusions

Dynamic programming can be used to solve optimal control problems by recursively evaluating a model for each state in each time step over a search horizon. Doing so requires a large number of evaluations of a model, but most of these computations can be done in parallel. The proposed FPGA architecture allows several parallel executions of a pipelined model which are used to compute a cost-to-go map. The architecture is scalable between a configurable number of execution units to enable a trade-off between performance and FPGA resource usage. The resulting hardware performs the optimization in less than 1.5% of the time of the original CPU version, and achieves performance that allows real-time use of the algorithm as opposed to running it off-line.

References

1. Bellman, R.: The theory of dynamic programming. Bull. Amer. Math. Soc. **60**(6), 503–516 (1954). https://doi.org/10.1090/s0002-9904-1954-09848-8
2. Bemporad, A.: Explicit model predictive control. In: Baillieul, J., Samad, T. (eds.) Encyclopedia of Systems and Control, pp. 1–9. Springer, London (2013). https://doi.org/10.1007/978-1-4471-5102-9_10-1
3. Bertsekas, D.P.: Dynamic Programming and Optimal Control. Athena Scientific, Belmont (1995)
4. Gersnoviez, A., Brox, M., Baturone, I.: High-speed and low-cost implementation of explicit model predictive controllers. IEEE Trans. Control Syst. Technol. **27**(2), 647–662 (2019). https://doi.org/10.1109/tcst.2017.2775187
5. Hellström, E., Ivarsson, M., Åslund, J., Nielsen, L.: Look-ahead control for heavy trucks to minimize trip time and fuel consumption. Control. Eng. Pract. **17**(2), 245–254 (2009). https://doi.org/10.1016/j.conengprac.2008.07.005
6. Hu, Y., Georgiou, P.: A study of the partitioned dynamic programming algorithm for genome comparison in FPGA. In: Proceedings of IEEE International Symposium on Circuits and Systems, pp. 1897–1900 (2013). https://doi.org/10.1109/ISCAS.2013.6572237
7. Ingole, D., Holaza, J., Takacs, B., Kvasnica, M.: FPGA-based explicit model predictive control for closed-loop control of intravenous anesthesia. In: Proceedings of International Conference on Process Control. IEEE (2015). https://doi.org/10.1109/pc.2015.7169936
8. McInerney, I., Constantinides, G.A., Kerrigan, E.C.: A survey of the implementation of linear model predictive control on FPGAs. IFAC-PapersOnLine **51**(20), 381–387 (2018). https://doi.org/10.1016/j.ifacol.2018.11.063
9. Moussalli, R., Salloum, M., Najjar, W., Tsotras, V.J.: Massively parallel XML twig filtering using dynamic programming on FPGAs. In: Proceedings of IEEE International Conference on Data Engineering. IEEE (2011). https://doi.org/10.1109/icde.2011.5767899
10. Natarajan, S., et al.: ReneGENE-DP: accelerated parallel dynamic programming for genome informatics. In: Proceedings of IEEE International Conference on Electronics, Computing and Communication Technologies. IEEE (2018). https://doi.org/10.1109/conecct.2018.8482378
11. Pérez, L.V., Bossio, G.R., Moitre, D., García, G.O.: Optimization of power management in an hybrid electric vehicle using dynamic programming. Math. Comput. Simul. **73**(1–4), 244–254 (2006). https://doi.org/10.1016/j.matcom.2006.06.016
12. Settle, S.O.: High-performance dynamic programming on FPGAs with OpenCL. In: Proceedings of IEEE High Performance Extreme Computing Conference, pp. 1–6 (2013)
13. Skarman, F., Gustafsson, O.: Spade: an expression-based HDL with pipelines. In: Proceedings of Workshop on Open-Source Design Automation (2023)
14. Skarman, F., Gustafsson, O., Jung, D., Krysander, M.: Acceleration of simulation models through automatic conversion to FPGA hardware. In: Proceedings of International Conference on Field-Programmable Logic and Applications. IEEE (2020). https://doi.org/10.1109/fpl50879.2020.00068

15. Skarman, F., Gustafsson, O., Jung, D., Krysander, M.: A tool to enable FPGA-accelerated dynamic programming for energy management of hybrid electric vehicles. IFAC-PapersOnLine **53**(2), 15104–15109 (2020). https://doi.org/10.1016/j.ifacol.2020.12.2033

16. Skarman, F., Sörnäs, G., Thörnros, E., Gustafsson, O.: Spade. https://doi.org/10.5281/zenodo.7713114

17. Tucci, L.D., O'Brien, K., Blott, M., Santambrogio, M.D.: Architectural optimizations for high performance and energy efficient Smith-Waterman implementation on FPGAs using OpenCL. In: Design, Automation & Test in Europe Conference & Exhibition. IEEE (2017). https://doi.org/10.23919/date.2017.7927082

18. Wang, R., Lukic, S.M.: Dynamic programming technique in hybrid electric vehicle optimization. In: Proceedings of IEEE International Electric Vehicle Conference, pp. 1–8. IEEE (2012). https://doi.org/10.1109/IEVC.2012.6183284

19. Zhu, Z., et al.: A GPU implementation of a look-ahead optimal controller for eco-driving based on dynamic programming. In: Proceedings of European Control Conference. IEEE (2021). https://doi.org/10.23919/ecc54610.2021.9655197

Optimization Approaches for Efficient Deployment of Signal and Image Processing Applications

Wordlength Optimization for Custom Floating-Point Systems

Quentin Milot[1], Mickaël Dardaillon[1(✉)], Justine Bonnot[2],
and Daniel Menard[1]

[1] Univ Rennes, INSA Rennes, CNRS, IETR - UMR 6164, 35000 Rennes, France
{quentin.milot,mickael.dardaillon,daniel.menard}@insa-rennes.fr
[2] WedoLow, 35510 Cesson-Sévigné, France
jbonnot@wedolow.com

Abstract. Algorithm complexity and problem size explosion during the last few years brought new challenges for hardware implementation of data-oriented applications. In particular, they have caused a large increase in memory, execution time, and power needed. One solution to address those challenges is to reduce data sizes. Custom floating-point is a good candidate that brings a large dynamic range in a compact representation. Thanks to improvement in floating-point units, it is more and more explored as an alternative to fixed-point.

This paper proposes an automatic optimization flow to optimize floating-point wordlength for a given quality metric. This flow is generic and can be used for any C or C++ algorithm. Different strategies are proposed to optimize the exponent and mantissa wordlengths. These strategies take advantages of analyzing the data dynamic range to reduce the optimization time. The obtained results show that better implementation cost can be obtained compared to 16-bit floating-point for a same quality.

Keywords: Custom floating-point · Wordlength · Optimization

1 Introduction

During the last years, with the technological progress in the semiconductor industry, digital platforms have integrated more and more transistors and have become faster and more energy efficient. With this evolution, numerous smarter systems are designed based on data-oriented applications. This technological progress leads to a significant improvement of the logical elements through the reduction of latency and dynamic energy. The energy to transport the data can now be higher than the energy to process the data by up to two to four orders of magnitude [4]. Memory access and data transportation became the new bottleneck for computation optimization. One way to face this issue is to optimize the data wordlength [6].

Two finite precision arithmetics are mainly used inside digital platforms to implement data-oriented applications: fixed-point and floating-point. The fixed-point arithmetic was favored in energy-efficient systems thanks to its lower cost.

Even if shift operations must be inserted between operations for scaling or aligning data, the processing part of fixed-point is less complex in terms of logical elements compared to floating-point. For fixed-point arithmetic, the quantization step (distance between two consecutive values) is constant due to the fixed implicit scaling factor. For floating-point arithmetic, the quantization step is adapted to the value to represent thanks to the explicit scaling factor embedded in the floating-point data. Thus, floating-point arithmetic can better compress data than fixed-point *i.e.* for a similar quantization error, fewer bits are required for encoding a floating-point data than for a fixed-point data.

However, with the reduction of floating-point's computing energy, especially compared to memory access energy [6], custom floating-point data types becomes a new contender for energy-efficient systems integrating data-oriented applications. In contrast to 64, 32 and 16-bit standard floating-point data types, customizing floating-point data requires an in-depth knowledge of the data dynamic range and application accuracy to adapt the wordlength to the need of each data. To obtain an optimized cost, the wordlength of the mantissa and the exponent must be adapted for each data inside an application. To face the complexity of modern applications, and to limit the development time, automatic frameworks to convert an application using standard floating-point data types to custom floating-point data types are mandatory.

Different methodologies and frameworks [2] have been proposed for fixed-point conversion. This conversion is composed of two main steps. First, the number of bits for the integer part is determined from the dynamic range analysis. Secondly, the number of bits for the fractional part is determined. The total number of bits is optimized such as the implementation cost is minimized subject to a quality criterion on the application output. In floating-point arithmetic, the problem is more complex because the accuracy depends on the mantissa wordlength but also of the exponent wordlength. Wordlength optimization for custom floating-point arithmetic has been proposed for specific applications like deep learning [8,10,13]. Generic approaches have been proposed for custom floating-point system design [3,12]. However, the strategy to optimize the wordlength of both the exponent and mantissa is not considered in these approaches.

In this paper, a new method for custom floating-point refinement is proposed. Different strategies are proposed to optimize the exponent and mantissa wordlengths. These strategies take advantage of analyzing the data dynamic range to reduce the optimization time. The obtained results show that better implementation costs can be obtained compared to standard 16-bit floating-point for a same quality.

The rest of the paper is organized as follows. Section 2 presents previous works on finite precision conversion. The proposed finite precision conversion flow is then described in Sect. 3. The determination of the exponent wordlength from dynamic range analysis is presented in Sect. 4. The optimization strategies specified for custom floating-point are detailed in Sect. 5. Section 6 presents the experiment results on different use cases.

2   Background and State of the Art

2.1   Finite Precision Arithmetic

Two finite precision arithmetics are mainly used to implement data-oriented applications: fixed-point and floating-point. Fixed-point arithmetic allows representing real values with the help of integer arithmetic. A fixed-point value x_{FxPt} is composed of two parts corresponding to the integer and the fractional part. x_{FxPt} is obtained by multiplying an integer x_{int} by a constant scaling factor 2^{-n}, where n corresponds to the fractional part wordlength.

Contrary to fixed-point arithmetic, for which the scaling factor is implicit and not embedded in the data, in floating-point arithmetic, the scaling factor is explicit and defined through the exponent field e. A normalized floating-point value x_{FlPt} is composed of three parts corresponding to the sign s, the exponent e and the mantissa m. This exponent e is embedded in the encoded data and allows selecting the range by representing the $\log_2$ of the number x to represent. Thus, with floating-point data types, very small and very high values can be represented with a good relative accuracy. The mantissa part m refines the value inside the considered range.

To avoid multiple representations of a given value and to allow interoperability between digital platforms, standards, like the IEEE 754 norm have been proposed for defining floating-point data type for 64, 32 and 16 bits. A normalized floating-point data type is under discussion for 8-bit data [7,11]. This standard defines the exponent and mantissa wordlengths, w_E and w_M respectively [5].

The exponent e_c embedded in the floating-point data is an unsigned integer value. To consider negative exponents to represent small values, a bias Δ equal to half of the exponent range is defined with the following expression:

$$e = e_c - \Delta \qquad \text{with} \qquad \Delta = 2^{w_E - 1} - 1$$

where w_E corresponds to the number of bits to represent the exponent. The minimal and maximal exponent values are used to represent specific values (0, denormalized values, $\pm\infty$, NaN).

In hardware design, like for fixed-point arithmetic, optimizing the data wordlength and using its own wordlength for each data will reduce the implementation cost (area, energy, latency). For custom floating-point data, reducing the exponent wordlength will impact the range of the data that can be represented. Reducing the mantissa wordlength will impact the accuracy [10]. The challenge is to provide automated design tools to convert a C source code into a C++ code using custom floating-point data types. The data wordlength search space must be explored efficiently in order to reduce the implementation cost for a maximal application quality degradation.

2.2   State of the Art

The challenge of creating design tools for custom floating-point data type has been explored through multiple point of view in the literature.

Domain-Specific Quantization. Domain-specific quantization takes advantage of domain particularities to improve quantization. Neural network quantization is a good example, with current interest in machine learning driving a lot of works in both quantization-aware training and post-training quantization. In quantization-aware training, data wordlengths are added as parameters to optimize during training [8]. It improves quantization robustness by taking into account quantization noise during training, but the method is specific to machine-learning. Post-training quantization is more generic as it can be applied outside of the domain of machine learning. It relies on the algorithm designer to set an objective. These approaches usually optimize a wordlength per layer, and use standard floating-point data types based on the targeted architecture [13]. In [10], custom floating-point data types are used. The exponent and mantissa wordlength are optimized.

Bitsize. [3] The Bitsize framework aims at optimizing the wordlength for fixed-point and floating-point data types. This framework uses an analytical approach for quantization error evaluation. The error model is efficient in terms of error evaluation time but limited to smooth operations. An operation is considered as smooth if the output is a continuous and differentiable function of the inputs.

Minibit+. [12] Minibit+ is an increment of Minibit approach [9] dedicated to fixed-point optimization. The transformation focuses on non-uniform optimization and floating-point optimization. The flow is based on a C++ program input. It starts with a range analysis using affine arithmetic. It then performs a coarse-grained accuracy analysis followed by a fine-grained accuracy analysis to determine the data wordlength. The main limitation is its lack of support for control structures such as conditions and loops.

In the Bitsize and Minibit+, the strategy to optimize the wordlength of both the exponent and mantissa is not detailed.

3   Finite Precision Conversion Flow

The proposed framework aims at converting automatically a C or C++ code with standard floating-point data types (float or double) to a C++ source code with custom floating-point data types for which the data wordlength has been optimized according to an application quality criterion. This output source code can then be used with high-level synthesis tools such as Vitis or Catapult for hardware implementation.

To optimize applications described with C/C++ floating-point code, the proposed framework in Fig. 1 starts with a source-to-source compiler [15]. This step analyzes the application source code to determine which variables in the application source code may have their precision optimized *i.e.*, all the variables involved in the application output computation. This first step also determines if some data types are set up by other ones, so as to reduce the design space to explore. Let $\mathcal{V}$ be a set grouping the N_v variables v_i to be optimized. At the output of this compiling step, the application source code is generated with

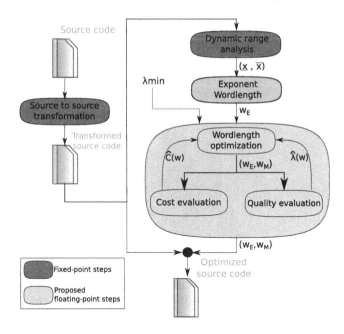

Fig. 1. Proposed framework for custom floating refinement.

generic data types that will be modified during the optimization process. This version of the application is then automatically instrumented by the framework to derive information on the variables belonging to the set $\mathcal{V}$.

The next step is a dynamic range analysis of each variable belonging to the set $\mathcal{V}$. Ranges are extracted based on a simulation running the testbench provided by the developer. The generic data types introduced in the first step are instrumented to collect all the values taken by a variable v during the source code execution and to build the histogram of $\log_2(v)$. To illustrate, an example of a histogram is given in Fig. 2.

4   Exponent Wordlength Determination

The dynamic range associated with standard floating-point data types is predefined and depends on the exponent wordlength. Since this range is not custom for standard floating-point data types, several bits in the exponent field can be left unused, leading to supplementary implementation cost.

To obtain a suitable exponent wordlength, the data range needs to match with the representation for any mantissa wordlength w_M. A floating-point data having w_E bits for the exponent fields with standard bias Δ allows representing values in the range indicated in Eq. 1 in the extreme case of mantissa wordlength $w_M = 0$.

$$[-2^{2^{w_E-1}-1}; -2^{-(2^{w_E-1}-2)}] \cup [2^{-(2^{w_E-1}-2)}; 2^{2^{w_E-1}-1}] \tag{1}$$

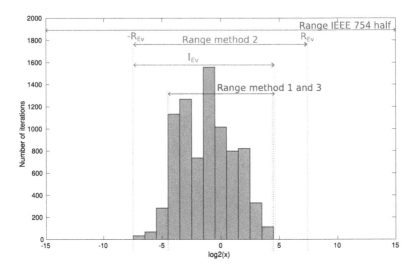

Fig. 2. Example of an histogram obtained from dynamic range analysis. Ranges for different floating-point data types are illustrated based on methods presented in Sect. 5.2.

Let $\underline{v}$ and $\overline{v}$ be respectively the minimal and maximal values taken by the variable $|v|$ without considering the value 0. The exponent range I_{E_v} represents the values that the exponent needs to reach to handle all the values of v. Since the exponent part needs to be symmetrical with standard bias Δ and reach both values of I_{E_v}, the maximal range distance R_{E_v} determines the maximal wordlength $w_E^{\max}$. The maximal wordlength $w_E^{\max}$ needs to represent the value R_{E_v} and $-R_{E_v}$ and can be determined with the following expression:

$$I_{E_v} = \left[\log_2(\underline{v}), \log_2(\overline{v})\right], R_{E_v} = \max\left(|I_{E_v}|\right), w_E^{\max} = \lceil\log_2\left(R_{E_v}\right)\rceil + 1 \quad (2)$$

Let $\mathbf{w}_E^{\max} = \left[w_{E_0}^{\max}, \dots w_{E_i}^{\max} \dots w_{E_{N_v-1}}^{\max}\right]$ be a N_v-length vector storing for each variable v_i of $\mathcal{V}$, the maximal number of bits $w_{E_i}^{\max}$ for the exponent.

5   Wordlength Optimization for Custom Floating-Point

5.1   Wordlength Optimization Problem Definition

Worldlength optimization aims at obtaining a minimal cost C satisfying a user-defined quality constraint λ_{min}.

$$\min_{\mathbf{w}}(\hat{C}(\mathbf{w})) \quad \text{subject to} \quad \hat{\lambda}(\mathbf{w}) \geq \lambda_{min} \quad (3)$$

With $\mathbf{w}$ being the concatenation of the vectors $\mathbf{w_E}$ and $\mathbf{w_M}$ representing the exponent and mantissa wordlengths respectively for each variable of the set $\mathcal{V}$.

Cost Evaluation. Different metrics like architecture area, energy consumption or memory space can be considered for the implementation cost C. For this study, an energy consumption and a memory space models are considered.

The dynamic energy of an operation depends on its type (addition, multiplication, memory transfer, ...) and the input and output wordlengths. For a given technology node for ASIC or for a given chip for FPGA, a library is obtained from a characterization process in which the operation costs are evaluated for different combinations of exponent and mantissa wordlengths. The energy consumption of the complete system is the sum of all the operation costs used in the system. For this study, a library for 28nm FDSOI technology is used [1]. This library provides the implementation cost for arithmetic operations but not for memory transfer, limiting the cost model to processing, excluding on-chip and off-chip memory. The memory cost aims at estimating the memory space to store the application variables. To do so, the number of allocations of a variable is determined and multiplied to the total wordlength of the variable.

Quality Evaluation. Simulation-based and analytical approaches can be considered to evaluate the quality metric λ at the output of an application according to the data wordlength $\mathbf{w}$. Analytical approaches aim at providing a mathematical expression of the quality metric. With this mathematical expression, the quality evaluation time is low but the supported applications are limited to those having only smooth operations.

Simulation-based techniques are more generic and support any application. The quality metric is statistically evaluated from data collected with a Monte-Carlo simulation carried out on the testbench input data. The source code is modified to integrate the custom floating-point data types. In this work, ac_type [14] are used. The confidence in the statistical estimation of the quality metric depends on the number of input samples used for the Monte-Carlo simulation. Thus, a huge number of samples can be required, leading to high quality evaluation time.

Optimization Algorithm. The space to explore being a subspace of natural numbers, the optimization needs to be heuristic. Numerous heuristic methods have been proposed to optimize wordlengths and especially in the context of fixed-point systems [2].

In this paper, the *min+1* algorithm [2] has been adapted for wordlength optimization problem in the context of custom floating-point. This algorithm is composed of two steps. First, an initial solution is found to start the greedy algorithm deployed in the second step. All the variables of vector $\mathbf{w}$ are set to their maximal value and one variable $\mathbf{w}_i$ is decreased until the quality constraint is no more fulfilled. The minimal value fulfilling the constraint is recorded in the vector $\mathbf{w}_{\min}$. This process is repeated for each variable of vector $\mathbf{w}$. In the second step, a mildest-ascent greedy algorithm is applied to increment by one bit one of the variables of vector $\mathbf{w}$ at each iteration. This second step starts with the initial solution $\mathbf{w}_{\min}$, for which the quality constraint is not fulfilled. At each iteration a gradient is computed for each variable to find the best direction *i.e.* the variable for which a one-bit wordlength increment lead to the best quality improvement. The optimization algorithm stops when the quality constraint is fulfilled.

This optimization method needs to test numerous configurations. Let n_C be the number of tested configurations to obtain the final solution. Let n_S be the number of input samples required to obtain a quality evaluation with a given confidence interval. The time t_{Opt} required for this global optimization process can be determined with Eq. 4. In this equation, t_S and t_{comp} represent the execution time for one sample and the compilation time of the modified source code to evaluate the quality.

$$t_{Opt} = n_C * (t_{comp} + n_S * t_S) \tag{4}$$

5.2   Global Wordlength Optimization Strategies

In this section, the proposed strategies to optimize the wordlength of both the exponent and mantissa are detailed. As shown in Eq. 4, the optimization time is proportional to the number of tested configurations. This latter depends on the optimization search space. The search space is linked to three parameters, N_v the number of variables to optimize, $\mathbb{E}_r \doteq \{\mathbf{x} \in \mathbb{N}^{N_v} : 1 \leq x_i \leq w_{E_i}^{\max}\}$ the search space for exponent wordlength and $\mathbb{M} \doteq \{\mathbf{x} \in \mathbb{N}^{N_v} : 1 \leq x_i \leq W_m\}$ the search space for the mantissa wordlength W_m represent the mantissa wordlength of the reference type (24 for single precision, 53 for double precision) and $w_{E_i}^{\max}$ the maximal number of bits required for the ith variable according to the range analysis.

Exponent and Mantissa Wordlength Simultaneous Optimization. This approach optimizes the wordlengths of the exponent and mantissa simultaneously. The maximal exponent wordlength is defined in vector $\mathbf{w}_E^{\max}$ computed with Eq. 2 with the help of the range analysis step. The variable for the optimization process is a $2.N_v$-length vector $\mathbf{w}$ composed of the exponent wordlength w_{E_i} and the mantissa wordlength w_{M_i} for each variable v_i of the set $\mathbb{V}$. The domain to explore is then $\mathbb{E}_r * \mathbb{M}$. The number of elements n_p of the search space can be expressed with the following expression:

$$n_p = M^{N_v} * \prod_{i=0}^{N_v} w_{E_i}^{\max} \tag{5}$$

By considering the exponent wordlength as a variable to optimize, compared to the approaches described in following sections , this approach provides more degrees of freedom to find an optimized solution. Indeed, values with a small exponent can potentially be approximated without degrading too much the application quality λ. Furthermore, for these values, the unnormalized representation can help to represent a part of the smallest values. To optimize this unnormalized representation, both the mantissa and exponent need to be taken into account.

The use of the reduced space $\mathbb{E}_r$, instead of a space composed of the maximal exponent wordlength ,allows reducing the optimization time. Indeed, the values higher than $w_{E_i}^{\max}$ will lead to the same quality results for a superior cost and are irrelevant. Nevertheless, this approach leads to the largest number of possibilities and thus tends to the highest optimization time of the three proposed methods.

Mantissa Wordlength only Optimization. This approach aims at optimizing only the mantissa wordlength. The exponent wordlength is set to the value obtained with Eq. 2 ($\mathbf{w}_e = \mathbf{w}_e^{\max}$). The variable for the optimization process is a N_v-length vector $\mathbf{w}$ composed only of the mantissa wordlength w_{M_i} for each variable v_i of the set $\mathbb{V}$. This a priori information allows limiting the search space to the set $\mathbb{M}$. Compared to the first approach, the number of variables in the optimization process has been divided by two. Thus, the number of tested configurations during the optimization process and the optimization time are reduced. The number of elements n_p of the search space can be expressed with the following expression:

$$n_p = M^{N_v} \tag{6}$$

Nevertheless, this approach can lead to a solution less optimized compared to the one obtained with the first approach. This approach does not explore the potential benefit of reducing slightly the exponent wordlentgh in relation to the value obtained with Eq. 2.

Exponent and Mantissa Wordlength Sequential Optimization. This approach is a middle ground between the other two. The optimization problem is divided into two optimization processes carried-out one after the other. First the exponent wordlengths are optimized and then the mantissa. The variable for the first optimization process is a N_v-length vector $\mathbf{w}$ composed of the exponent wordlength w_{E_i} for each variable v_i of the set $\mathbb{V}$. The variable for the second optimization process is a N_v-length vector $\mathbf{w}$ composed of the mantissa wordlength w_{M_i} for each variable v_i of the set $\mathbb{V}$. The optimization is performed on a subset of the search space $\mathbb{E}_r * \mathbb{M}$. The number of elements n_p of the search space can be expressed with the following expression:

$$n_p = M^{N_v} + \prod_{i=0}^{N_v} w_{E_i}^{\max} \tag{7}$$

Since this approach is a middle ground between the two other methods in terms of exploration space, the number of configurations required tends to be a middle ground. However, optimizing with a sequential optimization algorithm such as *min+1* can lead for high λ_{min} to the exploration of more configurations.

This solution brings a new hyperparameter to the optimization process. The quality degradation due to finite precision must be budgeted between the two independent optimization processes. In this work, the goal $\lambda_{E_{min}}$ for the first optimization process was set to equal the quality of the original system.

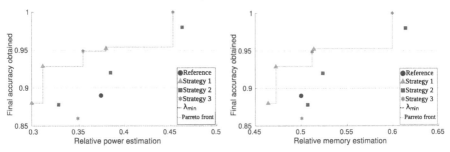

(a) Pareto chart of relative energy consumption (processing part) and final accuracy obtained according to the strategy and the minimal quality λ_{min}.

(b) Pareto chart of relative memory space and final accuracy obtained according to the strategy and the minimal quality λ_{min}.

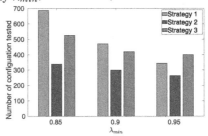

(c) Number of configurations tested for each optimization process according to the strategy and λ_{min}.

Fig. 3. Results for the squeezenet application.

6   Experiments and Results

The results are presented for two applications corresponding to a Linear Time-Invariant (LTI) system and a Convolutional Neural Network (CNN). The LTI system is an Infinite Impulse Response filter composed of four cascaded biquad cells. The optimization process is defined with $N_v = 7$. The quality evaluation metric is based on the Signal to Quantization Noise Ratio (SQNR), considering as a reference the single-precision floating-point data types (FP32). The second application is the Squeezenet CNN used for image classification. The network is composed of 14 layers. The optimization process is defined with $N_v = 15$. The CNN structure is complex due to pooling layers that add unsmooth operations. The quality evaluation metric is based on the top 1 accuracy, considering as a reference the FP32 implementation.

Firstly, the efficiency of our approach for finding an efficient solution in terms of tradeoff between cost and quality is evaluated for different quality metric constraints. For the Squeezenet, the tradeoff between the implementation cost and the quality metric is depicted in Fig. 3a and Fig. 3b respectively for the energy consumption and for the memory size. The x-axis represents the implementation cost normalized with the cost obtained for the reference solution with FP32. The y-axis represents the quality metric. The results are presented with the three proposed strategies and for three values of λ_{min}. The Pareto front is shown with the dotted line. These solutions can be compared with the one obtained for the IEEE 754 half-precision floating-point data type (FP16). These results show that the solutions located in the Pareto front and obtained with our approach are better than the FP16 solution. For a similar quality metric value, a difference of 7% and 3% for the relative energy consumption are measured. Indeed, FP16 is a generic data type that can not take advantage of the specificities and the limited variable dynamic range of the considered application. Moreover, customizing mantissa and exponent wordlength allows obtaining more flexibility in terms of cost and quality compared to standard data types.

When comparing the three proposed strategies, the second strategy is always less efficient than the two others. This shows that reducing the exponent wordlength compared to $w_{E_i}^{max}$ provides gain. Decreasing the exponent wordlength of one or two bits can be acceptable in terms of quality degradation. This is illustrated in Fig. 2, where the small values located in the left distribution tail will be approximated with unnormalized numbers. Even, a reduction of one bit is not negligible when the exponent wordlength is small.

The results obtained for the IIR filter are summarized in Table 1. The solution with FP16 and our solutions obtained for $\lambda_{min} = 50\ dB$ are close. The dynamic range of the variables inside the filter is very high and the five bits of the exponent are required to represent these values. As for Squeezenet, the second strategy always leads to a more conservative solution. By considering the two applications, both the first and third strategies can provide the best solution.

Table 1. cIIR filter for different strategies and λ_{min}.

	Strategy	Relative power estimation	Relative memory estimation	SQNR obtained (db)	Number of configuration tested
λ_{min}	FP16	0.40	0.50	51.26	.
50	1	0.39	0.32	50.35	233
	2	0.42	0.45	51.18	154
	3	0.39	0.33	50.28	185
70	1	0.51	0.42	70.72	197
	2	0.54	0.55	70.10	125
	3	0.50	0.44	70.09	156
90	1	0.86	0.74	90.14	749
	2	0.87	0.81	90.956	112
	3	0.64	0.56	90.13	161

The number of configurations n_C tested during the optimization process are depicted in Fig. 3c for the Squeezenet and summarized in Table 1 for the IIR filter. By optimizing only the mantissa wordlength the second strategy always requires testing fewer configurations compared to the two others. Then, by using a sequential process, the third solution always requires half as much configurations compared to the first solution.

7   Conclusion

This paper proposes a new method for custom floating-point refinement. Three strategies were established to optimize the exponent and mantissa wordlengths according to a user-defined quality constraint. These strategies are based on data dynamic range analysis to adapt the exponent wordlength. The results show an improvement in terms of power and memory consumption compared to a generic solution with FP16 data types. Our approaches allow obtaining different trade-offs between cost and quality.

References

1. Barrois, B., et al.: Customizing fixed-point and floating-point arithmetic - a case study in K-means clustering. In: International Workshop on Signal Processing Systems (SiPS), pp. 1–6 (2017). https://doi.org/10.1109/SiPS.2017.8109980
2. Caffarena, G.: Wordlength optimization of fixed-point algorithms. In: Approximate Computing Techniques: From Component-to Application-Level, pp. 261–284. Springer (2022). https://doi.org/10.1007/978-3-030-94705-7_9
3. Gaffar, A., et al.: Unifying bit-width optimisation for fixed-point and floating-point designs. In: Symposium on Field-Programmable Custom Computing Machines, pp. 79–88 (2004). https://doi.org/10.1109/FCCM.2004.59
4. Horowitz, M.: Computing's energy problem (and what we can do about it). In: International Solid-State Circuits Conference (ISSCC) (2014)
5. IEEE Computer Society: IEEE Standard for Floating-Point Arithmetic (2008)
6. Jouppi, N., et al.: Ten lessons from three generations shaped google's TPUv4i : industrial product. In: International Symposium on Computer Architecture (ISCA), pp. 1–14. Valencia, Spain (2021). https://doi.org/10.1109/ISCA52012.2021.00010
7. Kuzmin, A., Van Baalen, M., Ren, Y., Nagel, M., Peters, J., Blankevoort, T.: Fp8 quantization: the power of the exponent. Adv. Neural. Inf. Process. Syst. **35**, 14651–14662 (2022)
8. Kwak, J., et al.: Quantization aware training with order strategy for CNN. In: International Conference on Consumer Electronics-Asia, pp. 1–3 (2022). https://doi.org/10.1109/ICCE-Asia57006.2022.9954693
9. Lee, D., et al.: Minibit: bit-width optimization via affine arithmetic. In: Conference on Design automation (DAC), p. 837 (2005). https://doi.org/10.1145/1065579.1065799
10. Liu, F., et al.: Improving neural network efficiency via post-training quantization with adaptive floating-point. In: International Conference on Computer Vision (ICCV), pp. 5261–5270. Montreal, Canada (2021). https://doi.org/10.1109/ICCV48922.2021.00523
11. Micikevicius, P., et al.: FP8 Formats for Deep Learning (2022)

12. Osborne, W., et al.: Automatic accuracy-guaranteed bit-width optimization for fixed and floating-point systems. In: International Conference on Field Programmable Logic and Applications, pp. 617–620 (2007). https://doi.org/10.1109/FPL.2007.4380730
13. Shah, H., et al.: KD-Lib: A PyTorch library for Knowledge Distillation, Pruning and Quantization. ArXiv (Nov 2020). https://doi.org/10.48550/arXiv.2011.14691
14. SIEMENS EDA: Algorithmic C Datatypes reference manual (2022)
15. WedoLow: Software engineering. https://www.wedolow.com/

An Initial Framework for Prototyping Radio-Interferometric Imaging Pipelines

Sunrise Wang[1]([✉])(ID), Nicolas Gac[1](ID), Hugo Miomandre[2](ID),
Jean-Francois Nezan[2](ID), Karol Desnos[2](ID), and Francois Orieux[1](ID)

[1] CNRS, CentraleSupélec, Laboratoire des signaux et systèmes, Université
Paris-Saclay, 91190 Gif-sur-Yvette, France
`sunrise.wang@centralesupelec.fr,`
`{nicolas.gac,francois.orieux}@universite-paris-saclay.fr`
[2] Univ Rennes, INSA Rennes, CNRS, IETR – UMR 6164, 35000 Rennes, France
`{hugo.miomandre,jean-francois.nezan,karol.desnos}@insa-rennes.fr`

Abstract. Although large radio-telescope arrays allow us to observe
the celestial sphere with an unprecedented level of detail and sensitivity, additional antennas drastically increases the cost of processing and
storing their data, complicating the design of computing hardware. Our
overall goal is to provide a system, which we term SimSDP, to aid in
their design. It will achieve this by providing resource usage estimations
for some given imaging pipeline and hardware architecture, allowing for
more informed decisions when building the production systems. We lay
the groundworks in this paper by presenting and validating an initial
system that implements three different imaging pipelines. We find that
in most cases, our system is able to accurately estimate the scaling across
both algorithmic parameters as well as parallelization when compared to
measured data, with errors roughly in the 1–5% range, demonstrating
its ability to inform design decisions.

Keywords: Algorithm Design Space Exploration · SKAO ·
PREESM · Radio-Interferometry · Resource Estimation

1 Introduction

Radio-telescopes capture information of our skies within the radio spectrum,
allowing for the study of a host of otherwise invisible natural phenomena, with
some examples including unionized gas clouds, and low-frequency synchotron
radiation. Although many initial radio-telescopes were single-dish instruments,
modern advancements in the field of aperture synthesis allows us to instead use
antenna arrays with radio-interferometry. This is advantageous as it allows for
much larger apertures and higher sensitivities, allowing us to observe smaller
and fainter objects.

The currently under construction Square Kilometer Array (SKA) is an example of such a telescope, and on completion, will be the largest antenna array in

T. Dias and P. Busia (Eds.): DASIP 2024, LNCS 14622, pp. 56–67, 2024.
https://doi.org/10.1007/978-3-031-62874-0_5

the world. An issue with the SKA, as well as other large antenna arrays, is their computational cost, as the raw data obtained from the antennas is processed digitally. Thus, the more antennas, the larger the amount of data to be processed, with estimates for the SKA being around 0.4TB/s, leading to roughly 34.5 PBs per day. Storing this data for any significant period of time is prohibitively expensive, leading to strict time restrictions during processing. This, coupled with the fact that the type of science performed dictates the algorithmic pipeline, makes hardware and software design particularly challenging.

To aid in the above, a system that provides resource usage estimations for various radio-interferometric algorithmic pipelines and hardware architectures is desired. Unfortunately, as far as we are aware, none currently exists. The closest to this is DALiuGE [24], a graph-based execution framework targeted at radio-astronomy, and DASK [19], which has been used in radio-interferometry systems such as RASCIL [11]. However, from our understanding, these are primarily execution frameworks and do not provide easy resource usage estimations, particularly at scale.

It is our aim to introduce such a system, which will be termed SimSDP as it is targeted at the design of the SKA supercomputer, the Science Data Processor (SDP). This system will consist of two main parts. The first is responsible for both the algorithmic descriptions, as well estimating the performance for hardware at a local level, and the second aims to generalize the performance estimations across different hardware architectures.

This paper lays the groundwork for the former, by presenting a generic framework for radio-inteferometric algorithms, and providing three implementations of imaging pipelines within. We evaluate our framework's ability to aid in the design of the SDP by estimating the resource usage of these different pipelines across a variety of parameters and two levels of parallelization, and compare these against measured data to see if we can draw similar conclusions.

The main contributions of this paper are:

- A generic framework describing radio-interferometric algorithms, together with the full data-flow description and implementation of three different imaging pipelines,
- An evaluation of our framework's ability to aid in the design of the SDP, by comparing both estimated resource usages against measured across a variety of parameters and two parallelization levels,

The remainder of this paper is structured as follows: Sect. 2 provides a brief introduction to radio-interferometry as well as the generic pipeline, Sect. 3 discusses our implementation of this, and provides details on our implemented algorithms and how we perform our estimations, Sect. 4 discusses our results, and Sect. 5 concludes our work.

2 Radio-Interferometric Imaging

Radio-interferometers measure the sky using antenna arrays. Samples are produced by pairs of antennas in the array, termed baselines. Each sample, termed a visibility, is the instrumental response for some given time duration and electromagnetic frequency for a specific baseline. Visibilities can be defined with the Radio-Interferometric Measurement Equation (RIME) [21]:

$$V(u,v) = G_{uv} \int D_{uv}(l,m) \frac{1}{\sqrt{1-l^2-m^2}} I(l,m) e^{-2\pi i(ul+vm+w(\sqrt{1-l^2-m^2}-1))} \mathrm{d}l\mathrm{d}m \qquad (1)$$

where G denotes the direction independent effects, such as antenna gain, D denotes the direction dependent effects, such as Faraday rotation caused by the Earth's ionosphere, (u,v) refers to the difference between antenna positions in the frame of the Earth's rotation, (l,m) refers to spatial coordinates on the celestial sphere, and I is the true sky.

If the D and $e^{w(\sqrt{1-l^2-m^2}-1)}$ terms are ignored, Eq. 1 simplifies to

$$V(u,v) = G_{uv} \int I(l,m) e^{-2\pi i(ul+vm)} \mathrm{d}l\mathrm{d}m \qquad (2)$$

a 2-dimensional Fourier transform, allowing us to retrieve an image of the true sky through an inverse Fourier transform.

The image produced by the inversion of Eq. 2, termed the dirty image, contains artefacts caused both by the partial sampling of the Fourier domain, as antenna arrays are inherently sparse, as well as by the omission of the w and D terms. The former results in a convolution between the Fourier transform of the sampling pattern, i.e. the Point Spread Function (PSF), and the real sky, while the latter results in non-isoplanatic image-plane distortions. Radio-interferometric imaging aims primarily to correct these in order to produce an image use-able for scientific purposes.

The Radio-Inteferometric pipeline [5] employs a nested loop structure (Fig. 1). The outer i.e. major loop Δ calculates the difference $\delta\hat{i}_n$ between some estimated sky image $\hat{i}_n$ and the measured values v, with $\hat{i}_0$ typically being a blank image.

Although there are many possible approaches to Δ, the most common general method currently involves transforming $\hat{i}_n$ to the same domain as v using a degridding operator, which typically includes a Fourier transform and an extrapolation operator, performing the subtraction to obtain the difference in visibility space, and then using a gridding operator which includes an inverse Fourier transform together with an interpolation operator to obtain $\delta\hat{i}_n$.

In addition to computing the difference, the major loop is also responsible for correcting for the w and D terms. This can be achieved either in tandem with the gridding convolution stage $G^\dagger$ [3,9,22] or through various discretization methods [10,15,16]. There has also been work that employs a hybrid of the two strategies [6,25].

Finally, the major loop also allows for calibration of the direction independent effects G, which we don't show in Fig. 1 as it is not the focus of our work.

The inner i.e. minor loop Ψ is responsible for removing the convolution artefacts caused by the partial sampling of the Fourier domain from $\delta\hat{i}_n$, and then combining this information with $\hat{i}_n$ to produce an estimate $\hat{i}_{n+1}$ for the

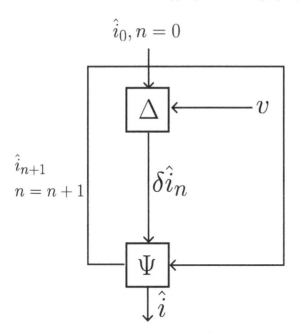

Fig. 1. The general framework for producing the final sky estimate $\hat{i}$ is iterative, where $\hat{i}_n$ is updated while taking into account the difference $\delta\hat{i}_n$ between it and the measurements v. The degridding-gridding step (Δ) is responsible for computing this, whereas the deconvolution step Ψ is responsible for generating the next estimate $\hat{i}_{n+1}$.

next major cycle iteration. This is achieved by employing a deconvolution algorithm, with some popular examples being CLEAN [12] and its variants [7,18,20], and convex optimization methods that regularize based on entropy [8] or sparsity [1,4,23].

As the deconvolution methods can be partial or contain errors in their estimation, the new estimate $\hat{i}_{n+1}$ is passed back to the major loop. This is done until $\delta\hat{i}_n$ is predominantly noise.

3   Implementation

We implement the generic structure defined in Fig. 1, as well as three different imaging algorithms using PREESM [17]. We find PREESM to be particularly well suited to our needs, as algorithms are defined in data-flow diagrams, which allows for easy definition of additional imaging algorithms. It also allows for automated parallelism, which we use for our different levels of parallelization. Finally, PREESM also allows for easy estimation of resource consumption when provided with the relevant data for each actor.

The generic top-level data-flow diagram provides a common interface for algorithms defined in Δ and Ψ i.e. degridding-gridding and deconvolution. This enables us to encompass a large number of different imaging pipelines, as most work in the domain tend to focus on either.

The rest of this section will look to provide a high-level overview on the algorithms we implemented for both Δ and Ψ, as well as how parallelization is performed. It also discusses how we perform our resource usage estimations. We provide our system and data in our on-line repository[1].

3.1   Algorithms for Δ and Ψ

We implement three different degridding-gridding algorithms (Δ) in our pipeline. The first is the Direct Fourier Transform (DFT) degridder, which transforms the image directly to the visibilities using:

$$V(u,v) = \sum_{l=0,m=0}^{L,M} I(l,m)e^{-2\pi i(ul+vm)} \tag{3}$$

As it is unnecessary to perform the equation for areas where I is zero, the overall complexity for this method is $\mathcal{O}(n_v n_s)$, where n_v is the total number of visibilities, and n_s is the number of sources.

Following this is the Fast-Fourier Transform (FFT) degridder, which first performs a Fast Fourier Transform on the image to obtain the gridded visibilities, which are then subsequently convolved onto the continuous visibility positions. This algorithm has a complexity of $\mathcal{O}(n_g^2 \log_2 n_g + n_v n_{dgk})$, where n_g is the support size of the grid, and n_{dgk} is the support size of the de-gridding kernel.

Both the DFT and FFT degridders also perform a gridding operation after the subtraction of visibilities, which has a complexity of $\mathcal{O}(n_g^2 \log_2 n_g + n_v n_{gk})$ where n_{gk} is the size of the gridding kernel.

The last implemented degridding-gridding algorithm is the more recent Grid to Grid (G2G) method by Monnier et al. [14]. This method performs degridding and gridding in a single step, and performs the subtraction in image space rather than visibility. The total complexity of this algorithm is $\mathcal{O}(2n_g^2 \log_2 n_g + 2n_v(n_{gk} + n_{dgk}))$, the same as the FFT degridder. However, this algorithm is more performant, as it employs a visibility simplification stage which diminishes the size of n_v. Furthermore, the subtraction is done in image space, which may be faster as the number of visibilities often dwarfs the number of pixels in the image.

We choose to focus on degridding-gridding (Δ) algorithms for this work, thus only implement a single deconvolution algorithm (Ψ), that being Högbom CLEAN. [12], which we selected for its simplicity. This algorithm iteratively finds the brightest source in the image, and subtracts the PSF from its position, terminating after a preset number of minor cycles n_m. The complexity of this algorithm is $\mathcal{O}(n_g^2 + n_m n_p)$, where n_p is the support of the PSF. As this algorithm finds a single point source per iteration, $n_s = n_m$, which is pertinent for the DFT degridder.

3.2   Parallelization

To parallelize Δ, we design the data-flow diagram to divide the visibilities based on some specified number of threads, grid or degrid each subset of visibilities independently, and then merge each thread's results. This manner of parellelization is natural for Δ as it requires looping through the visibilities. We don't parallelize Ψ as Högbom CLEAN has cross iteration data dependencies, drastically complicating process.

We create two different levels of parallelization with 1 and 4 cores. We set the data division parameter in our data-flow diagrams to not divide the visibilities in the former case, and divide them into 4 disjoint subsets in the latter.

3.3   Estimating Resource Utilization

We perform computation time and memory estimations for our system. For the former, PREESM requires time information for each actor, either in the form of an exact value, or some function dependent on the input data. We opt for the latter as it affords us the ability to predict how the algorithms scale.

We perform benchmarking on optimized versions of the code for each actor, while varying the number of visibilities, the image size, and the number of minor cycle iterations. We then fit polynomials of the appropriate degree to these, which we then transfer to PREESM. We average 30 samples for each benchmark data point.

We perform the benchmarking for actors that contain simple and near-optimal implementations directly in PREESM. For more complex actors, such as the G2G and FFT degridding-gridding algorithms, we profile the code of Monnier et al. [14].

The dataset used to generate these benchmarks is obtained from the SKAO simulated database repository. We provide full information on this in Sect. 4.1. In order to vary the number of visibilities, we either truncate or duplicate them. We opt for this over random generation because the antenna layout may have an effect on the performance of some algorithms, such as G2G.

The machine used to obtain the benchmarks is a Dell Precision 3551, with an Intel Core i7 10875H 10th generation processor and 64GB of RAM. Re-profiling and fitting is required for other hardware as our fitted models become invalid.

Unlike the computational time estimations, memory estimations in PREESM are much less involved, with PREESM simply reporting the maximum allocated inter-actor memory. It does not estimate intra-actor memory as it does not perform any code analysis, and primarily serves as a lower-bound estimation.

4   Results and Discussion

This section provides details on the simulated dataset that we use for our experiments, our framework validation, as well as our resource estimation results.

4.1   Simulated Database

The simulated dataset used for our validation and experiments is the same as the one used for our benchmarking, and is the small database from the SKA simulated dataset repository[2]. It is simulated to have a field of view of 1°C, with 33 point sources within. The array has a total of 512 antennas, leading to 130816 baselines. The visibilities were sampled from each baseline once every 30 s, for a total of 15 min, creating 3924480 total visibilities.

This dataset is single polarization single wavelength, which is well suited to the current limitations of our system, as it is currently unable to handle different polarizations, extended sources, and multiple frequency channels.

4.2   Pipeline Validation

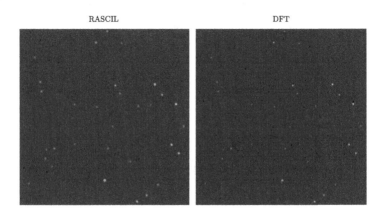

Fig. 2. Final deconvolved images of RASCIL and our DFT degridder-gridder. We can see that the sources appear in the same positions with similar energy ratios, showing that our pipeline functions as intended. Our other images are similar and available on our repository.

We validate our pipeline by performing a visual inspection of its reconstructed images to the ones produced from the well known imaging framework RASCIL [11]. Figure 2 shows the images produced both by RASCIL, and our DFT degridder. The other algorithms produce very similar images, and are available on our repository.

One can see that the general structure of the produced image is the same, with the point sources and their energy ratios being similar. The differences are primarily due to minor algorithmic discrepancies, such as parameters in the deconvolution algorithm.

---

[2] https://gitlab.com/ska-telescope/sim/sim-datasets.

4.3   Computing Time Estimations

We evaluate our computation time estimations by comparing them to the measured times of our pipelines obtained using the GNU time tool. We plot these against different numbers of grid cells, numbers of visibilities, and the numbers of minor cycles per major iteration.

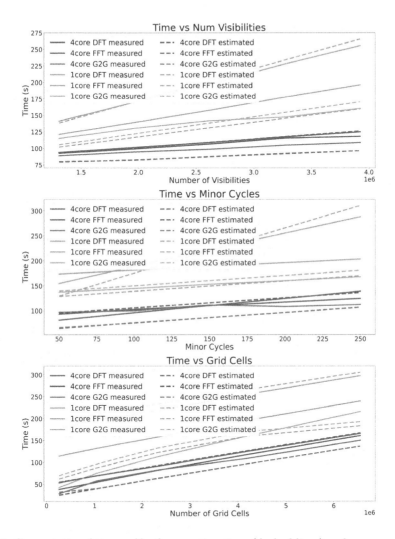

Fig. 3. Computational times of both our estimations (dashed lines) and measurements (solid lines) for our implemented methods across both levels of parallelization.

Figure 3 shows our results, where the dashed line denotes estimated times, the solid line denotes measured, the lighter colours denote the single core setup, and the darker denote the quad core.

We found our estimations to accurately predict the increase in time with increasing algorithm parameters, with most cases having an average error within the 1–5% range, allowing us to draw similar scaling conclusions to the measured data. For example, we can see that the DFT algorithm scales much poorer compared to the others with regards to the number of sources in the image, as well as the number of visibilities.

We also found that our framework allows us to draw valid conclusions with regards to scaling in terms of parallelization, with it accurately predicting in most cases the roughly $1.5 - 3\times$ speedup obtained when parallelizing across 4 cores.

Despite the positives, we found various areas in which the accuracy of our estimations was lacking. This can be seen for the FFT and G2G algorithms when varying the number of grid cells, and DFT in the cases where the minor cycles are varied. There are three main reasons for these poor estimates:

- With the exception of the DFT, the code benchmarked was much more optimized than the code measured;
- Our fitting functions introduce additional error;
- PREESM does not perform static analysis.

Although the former two points can be easily resolved by taking more accurate and a larger number of benchmark samples, the latter is a drawback to our current framework, which only performs static analysis. This is particularly evident in the results for G2G, which performs a visibility simplification step to dynamically reduce the number of visibilities needing to be processed.

4.4   Memory Usage Estimations

We evaluate the memory usage estimations by comparing them against the measured maximum resident set size obtained by the GNU time tool in verbose mode. We plot the memory usage against both the number of grid cells and the number of visibilities for both levels of parallelization. Unlike the computational time experiments, we do not vary the number of minor cycles as it does not have a major effect on memory usage.

Figure 4 shows the estimated and measured memory usage in our system. Much like the computation time estimations, our framework predicts well the increase when scaling the amount of data, with an error of 1–5% for all test cases. It also predicts the $1.1 - 1.5\times$ increase in memory usage when increasing parallelization.

These results are not surprising, as PREESM is responsible for allocating the inter-actor memory. However, it is unexpected that the measured memory usage is lower than the estimated. As mentioned in Sect. 3.3, the estimate should provide a lower-bound, which is in conflict with what we observed. We are still investigating possible explanations for this discrepancy, and will leave a more detailed analysis for future work.

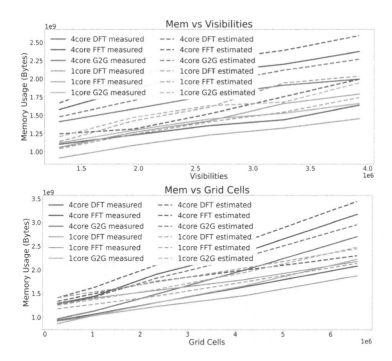

Fig. 4. Memory estimations (dashed lines) and measurements (solid lines) for the various algorithms implemented in PREESM plotted against both number of visibilities, and number of grid cells.

5   Conclusions and Future Work

In this paper, we lay the groundworks for SimSDP, a system aimed at aiding in the design of supercomputers processing data from large radio-telescope arrays by providing resource usage estimations at various scales and hardware architectures. We achieve this by performing a study on a preliminary system aimed at such, which includes three different algorithmic pipelines, and two levels of parallelization. We found that the estimations obtained from our framework allows us to draw similar scaling conclusions to the measured results. This gives impetus in continuing the design and implementation of SimSDP.

5.1   Drawbacks

One major drawback of our system is that it currently only performs static analysis, which may lead to it providing poor estimates in cases where algorithms perform dynamic reductions to the data being processed. Also in line with this, there are additional accuracy improvements that can be made to our estimations through more and better benchmarking data. Finally, our system currently does not handle many types of datasets, as it doesn't currently support objects containing extended sources (eg. nebulae), different polarizations, and multiple wavelengths.

5.2   Future Work

Much work still needs to be done for the realization of SimSDP. This includes: extending the current system to support different hardware architectures; implementing more algorithms for Δ and Ψ, which will allow for our system to support more types of datasets; introduce dynamic analysis by employing SPIDER [13]; estimating energy consumption; and comparing our system's estimations to measurements obtained from production systems [2,11,15].

Acknowledgements. This work was supported by DARK-ERA (ANR-20-CE46-0001-01)

References

1. Ammanouil, R., Ferrari, A., Mary, D., Ferrari, C., Loi, F.: A parallel and automatically tuned algorithm for multispectral image deconvolution. Mon. Not. R. Astron. Soc. **490**(1), 37–49 (2019). https://doi.org/10.1093/mnras/stz2193
2. Bean, B.E.A.: CASA, the Common Astronomy Software Applications for Radio Astronomy. Publicat. Astronomical Soc. Pacific **134**(1041), 114501 (2022). https://doi.org/10.1088/1538-3873/ac9642
3. Bhatnagar, S., Cornwell, T.J., Golap, K., Uson, J.M.: Correcting direction-dependent gains in the deconvolution of radio interferometric images. Astronomy Astrophys. **487**(1), 419–429 (2008). https://doi.org/10.1051/0004-6361:20079284
4. Carrillo, R.E., McEwen, J.D., Wiaux, Y.: PURIFY: a new approach to radio-interferometric imaging. Mon. Not. R. Astron. Soc. **439**(4), 3591–3604 (2014). https://doi.org/10.1093/mnras/stu202
5. Clark, B.: An efficient implementation of the algorithm 'CLEAN'. Astronomy Astrophys. **89**, 377 (1980)
6. Cornwell, T.J., Voronkov, M.A., Humphreys, B.: Wide field imaging for the square kilometre array, San Diego, California, USA, p. 85000L (Oct 2012). https://doi.org/10.1117/12.929336
7. Cornwell, T.J.: Multiscale clean deconvolution of radio synthesis images. IEEE J. Selected Topics Signal Process. **2**(5), 793–801 (2008). https://doi.org/10.1109/JSTSP.2008.2006388
8. Cornwell, T.J., Evans, K.F.: A simple maximum entropy deconvolution algorithm. Astronomy Astrophys. **143**, 77–83 (1985), (ISSN 0004-6361)
9. Cornwell, T.J., Golap, K., Bhatnagar, S.: The noncoplanar baselines effect in radio interferometry: The W-projection algorithm. IEEE J. Selected Topics Signal Process. **2**(5), 647–657 (2008). https://doi.org/10.1109/JSTSP.2008.2005290
10. Cornwell, T.J., Perley, R.A.: Radio-interferometric imaging of very large fields-the problem of non-coplanar arrays. Astronomy Astrophys. **261**, 353–364 (1992)
11. Cornwell, T.J., Wortmann, P., Nikolic, B., Wang, F., Stolyarov, V.: Radio astronomy simulation, calibration and imaging library (2020)
12. Högbom, J.A.: Aperture synthesis with a non-regular distribution of interferometer baselines. Astronomy Astrophys. Suppl. **15**, 417 (1974)

13. Miomandre, H., et al.: Demonstrating the SPIDER Runtime for Reconfigurable Dataflow Graphs Execution onto a DMA-based Manycore Processor. In: IEEE International Workshop on Signal Processing Systems (Oct 2017), poster
14. Monnier, N., Orieux, F., Gac, N., Tasse, C., Raffin, E., Guibert, D.: Fast Sky to Sky Interpolation for Radio Interferometric Imaging. In: 2022 IEEE International Conference on Image Processing (ICIP). pp. 1571–1575. IEEE, Bordeaux, France (Oct 2022). https://doi.org/10.1109/ICIP46576.2022.9897317
15. Offringa, A.R.E.A.: wsclean: an implementation of a fast, generic wide-field imager for radio astronomy. Monthly Notices Royal Astronomical Soc. **444**(1), 606–619 (2014). https://doi.org/10.1093/mnras/stu1368
16. Ord, S.M., et al., M.: Interferometric Imaging with the 32 Element Murchison Wide-Field Array. Publicat. Astronomical Soc. Pacific **122**(897), 1353–1366 (2010). https://doi.org/10.1086/657160
17. Pelcat, M., Desnos, K., Heulot, J., Guy, C., Nezan, J.F., Aridhi, S.: Preesm: A dataflow-based rapid prototyping framework for simplifying multicore DSP programming. In: 2014 6th European Embedded Design in Education and Research Conference (EDERC), pp. 36–40. IEEE, Milano, Italy (Sep 2014). https://doi.org/10.1109/EDERC.2014.6924354
18. Rau, U., Cornwell, T.J.: A multi-scale multi-frequency deconvolution algorithm for synthesis imaging in radio interferometry. Astronomy Astrophys. **532**, A71 (2011). https://doi.org/10.1051/0004-6361/201117104
19. Rocklin, M.: Dask: Parallel Computation with Blocked algorithms and Task Scheduling, Austin, Texas, pp. 126–132 (2015). https://doi.org/10.25080/Majora-7b98e3ed-013
20. Schwab, F.R.: Relaxing the isoplanatism assumption in self-calibration; applications to low-frequency radio interferometry. Astron. J. **89**, 1076 (1984). https://doi.org/10.1086/113605
21. Smirnov, O.M.: Revisiting the radio interferometer measurement equation: I. A full-sky Jones formalism. Astronomy Astrophy. **527**, A106 (2011). https://doi.org/10.1051/0004-6361/201016082
22. Van Der Tol, S., Veenboer, B., Offringa, A.R.: Image Domain Gridding: a fast method for convolutional resampling of visibilities. Astronomy Astrophys. **616**, A27 (2018). https://doi.org/10.1051/0004-6361/201832858
23. Wiaux, Y., Jacques, L., Puy, G., Scaife, A.M.M., Vandergheynst, P.: Compressed sensing imaging techniques for radio interferometry. Mon. Not. R. Astron. Soc. **395**(3), 1733–1742 (2009). https://doi.org/10.1111/j.1365-2966.2009.14665.x
24. Wu, C., et al.: DALiuGE: a graph execution framework for harnessing the astronomical data deluge. Astronomy Comput. **20**, 1–15 (2017). https://doi.org/10.1016/j.ascom.2017.03.007
25. Ye, H., Gull, S.F., Tan, S.M., Nikolic, B.: High accuracy wide-field imaging method in radio interferometry. Mon. Not. R. Astron. Soc. **510**(3), 4110–4125 (2022). https://doi.org/10.1093/mnras/stab3548

Scratchy: A Class of Adaptable Architectures with Software-Managed Communication for Edge Streaming Applications

Joseph W. Faye[1]([✉]), Naouel Haggui[1][ID], Florent Kermarrec[2],
Kevin J. M. Martin[3][ID], Shuvra Bhattacharyya[1,4][ID], Jean-François Nezan[1][ID],
and Maxime Pelcat[1][ID]

[1] IETR - UMR CNRS 6164, INSA Rennes, Rennes, France
{jofaye,nhaggui,jnezan,mpelcat}@insa-rennes.fr
[2] Enjoy-Digital, Landivisiau, France
florent@enjoy-digital.fr
[3] Université Bretagne Sud, Lab-STICC UMR 6285, Lorient, France
kevin.martin@univ-ubs.fr
[4] University of Maryland, College Park, USA
ssb@umd.edu

Abstract. Stream processing applications are becoming increasingly complex, requiring parallel and adaptable architectures under real-time constraints. Currently, selecting appropriate computing platforms for these applications is done manually through prototyping and benchmarking. To simplify this selection process, Dataflow (DF) modeling has been utilized to identify opportunities for parallelism. This approach utilizes the Algorithm Architecture "Adequation" (AAA) methodology to make efficient decisions at compile-time by considering data movement and scheduling needs in stream processing environments.

This paper presents a new architecture named "Scratchy", that is specially designed for stream processing applications. Scratchy is a multi-RISC-V architecture that features software-managed communication using scratchpad memories and customizable interconnect topologies. The architecture supports different topology options and is demonstrated using a 3-core Scratchy. The capabilities of the architecture are presented through a design space exploration that focuses on optimizing the topology for specific applications. It also highlights the low resource overhead of the architecture and quick synthesis time on an Intel MAX10.

Keywords: Dataflow Models of Computation · Streaming Applications · Hardware Architecture · Scratchpad Memories · System-On-Chip

© The Author(s), under exclusive license to Springer Nature Switzerland AG 2024
T. Dias and P. Busia (Eds.): DASIP 2024, LNCS 14622, pp. 68–79, 2024.
https://doi.org/10.1007/978-3-031-62874-0_6

1 Introduction

Edge computing brings proximity to cloud services, offering reduced latency, improved performance, and contextual awareness for pattern recognition and analysis and wearable devices [11]. Typically, edge devices process in-order data streams near their source with limited storage and computation.

DF Models of Computation (MoCs) provide semantics for modeling, analyzing, and optimizing embedded software for stream processing applications. These MoC define data dependencies between self-contained processing tasks called actors. Near-sensor streaming applications present complex control paths and require specialized tools and methodologies [1]. Achieving real-time performance in stream processing applications often requires leveraging application concurrency of multicore systems. The Symmetric Shared Memory Multiprocessor (SMP) multicore architecture facilitates the transition from sequential to parallel programs. It consists of multiple identical cores that are interconnected to a shared main memory via a Network on Chip (NoC), a bus, or a crossbar. These systems achieve speedups by dividing the application workload into concurrent tasks across multiple cores. However, the SMP architectures are built as Uniform Memory Access (UMA) machines where all data are supposed to be accessible with the same latency from all Processing Element (PE) of the System-on-a-Chip (SoC) and therefore do not easily exploit the coarse-grain application datapath. This explains the mainstream limitation of SMP to 16 cores.

This work aims to exploit the streaming nature of many applications for designing systems. In DF-described applications, synchronization costs arise from the granularity of data sharing. Multicore architectures, while beneficial of task-level parallelism, can restrain scalability due to increased thread synchronization [7]. Despite various mapping and scheduling solutions, hardware aspects have received less attention [7,14]. Hence, there is a need to develop adaptable architectures to improve the performance of streaming applications [9].

This paper introduces Scratchy, a multicore architecture designed for streaming applications. Scratchy favors RISC-V cores because of their open-source nature. One can create their cores from the basic ISA and easily integrate them into the LiteX tool. Scratchy utilizes Scratchpad Memories (SPMs) to ensure customizable inter-PE communication. A software library implements a queue definition and performs communication control at the software level to ensure inter-PE consistency. The paper also demonstrates Scratchy's capabilities through a Design Space Exploration (DSE) test case that aims to select the best Scratchy topology for two test applications described using Synchronous DataFlow (SDF).

The paper is structured as follows. Section 2 provides the background and state of the art. Section 3 presents an overview of the architecture. Section 4 describes the experimental setup of DSE, while Sect. 5 shows the results and a discussion. Finally, Sect. 6 concludes the article by discussing future work.

2   Background and State of the Art

High-performance embedded devices are crucial for edge and near-sensor computing, particularly for complex arithmetic calculations. Numerous multi-ARM solutions are currently available, such as the NVIDIA Jetson series, ASUS Tinker Board, and Kalray many-core DPU systems. However, selecting the right architecture for a specific application requires careful evaluation, which can be a complex task, and demands a significant investment of development time, and remains mostly manual. We propose an easy-to-generate, reconfigurable, and extensible architecture to address this challenge.

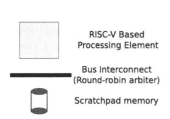

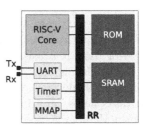

Fig. 1. Components of Scratchy   **Fig. 2.** RISC-V-Based PE

In computer architecture research, several frameworks are available for developing configurable SoCs. One such framework is OpenPiton [5], created to design, simulate, emulate, and build many-core cache-coherent and NoC based architectures. These architectures can span many computing subfields and run Linux [4]. Another Linux-capable architecture is BlackParrot, designed to be the default open-source, cache-coherent RV64GC multicore [20]. ESP [22] is another research platform that enables heterogeneous multicore soc designs using RISC-V. It is structured as a heterogeneous tile grid connected by a NoC. Other architectures, such as HERO [12], combine a Linux-capable host with a programmable many-core accelerator (PMCA). The first version uses an ARM Cortex-A multi-core processor host, while the second version has a LLVM-based heterogeneous compiler that supports OpenMP-coded applications. Another noteworthy research project in hardware design is the Chipyard SoC generation framework [2]. It combines various agile hardware design projects and is mainly based on the Rocket Chip, a chisel-based SoC generator [3]. Chipyard offers tools for developing target software workloads and simplifying hardware implementation with DF modeling.efforts [13,16,18,19]. Furthermore, DF can be combined with AAA methodology [18], allowing developers to express various forms of concurrency and exploit data and task parallelism. A notable feature of a static or quasistatic DF model is that it operates on a predetermined data path, making cache-based architectures less suitable for DF-modeled applications [8]. In the context of DF, the data is accessed only once, and the caching mechanism is not used. Moreover, using SPM for shared resources, communication can be defined and managed at the software level, resulting in lightweight communication. Communication can be customized to improve performance, and SPM can separate shared resource storage. This reduces contention overhead that is caused by synchronization and shared resource access.

This article proposes a class of architecture with three main components, as illustrated in Fig. 1, called Scratchy to provide an architecture tailored to the needs of stream processing applications modeled by DF MoCs. Scratchy enables inter-PE interconnection customization. The infrastructure enables multicore research by generating both the architecture and the multicore code.

3   Overview of the Scratchy Architecture

This section introduces Scratchy, an adaptable architecture for streaming applications modeled with DF MoC.

3.1   Architecture Components

Scratchy architecture is built using the LiteX [10] framework, which assists in creating FPGA cores/SoCs. It comprises three main components (PE, SPM, and bus) as shown in Fig. 1:

SPM: Connected to a unique bus and designed as double-port [6,21], each SPM stores data accessible by connected PEs. This setup supports zero-copy or First In, First Out (FIFO) queue-driven communication.

PE: A PEs are bus-centered cores that can include several IP components, as illustrated in Fig. 2. Each IP has its local ROM and RAM memories for the bootloader and user code storage. They feature an interrupt controller, timer, UART for debugging and code loading, and Memory Mapped Input/Outputss (MMIOs) for peripheral hardware control.

Bus Interconnect: A bus interconnects one or multiple PEs and a SPM. It allows connections and facilitates shared access to the SPM through master and slave interfaces. A round-robin arbiter schedules conflicting accesses.

3.2   Scratchy Infrastructure

This section outlines the RTL implementation and software support for Scratchy.

RTL Implementation: The Scratchy Register Transfer Level (RTL) implementation relies on the Migen library[1], a Python-based Florida Hardware Design

---

[1] https://m-labs.hk/gateware/migen/.

Language (FHDL) [15] that simplifies the design and simulation of digital systems by using combinatorial and synchronous statements instead of the traditional event-driven paradigm. Scratchy is based on LiteX and includes a generator that takes a Scratchy Configuration (SC) file in JSON syntax as input. This file stores the system-level description of the computing platform, including cores, SPMs, and interconnects. The generator then produces synthesizable code for a selected FPGA target.

Scratchy Multicore Compilation Infrastructure: The multicore code for Scratchy is generated from a hierarchical SDF description using the PREESM framework [19]. PREESM provides deadlock-free multicore code generation and simulation. The communication library provided by PREESM has been customized to support bare-metal targets. The Scratchy architecture generation and multicore compilation process is depicted in Fig. 3.

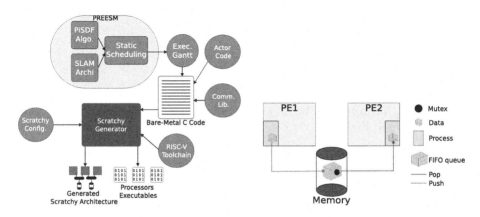

Fig. 3. Scratchy System Design Workflow

Fig. 4. FIFO-based communication library principle mechanism.

Middleware and Inter-PE Communication: The Scratchy infrastructure utilizes the LiteX SoC composer for software stack management, including bootloader loading and application launching. Key components include Compiler-rt[2] for low-level code generation, particularly for floating-point arithmetic on cores without an FPU, and Picolibc[3], which offers an API of the C library optimized for small embedded systems. Picolibc combines the Newlib[4] and AVR Libc[5] code, utilizing the Meson build system to compile across various platforms.

---

[2] https://compiler-rt.llvm.org/.
[3] https://keithp.com/picolibc/.
[4] https://sourceware.org/newlib/.
[5] https://www.nongnu.org/avr-libc/.

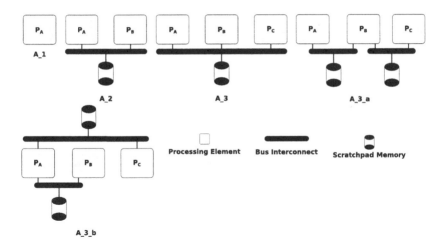

Fig. 5. Different generated scratchy architectures

Scratchy features a minimalist communication library for inter-core communication and synchronization. This includes:

Barrier Mechanism: A custom barrier mechanism using coherent SPM accesses, employing a *Centralized Barrier* algorithm [17] to manage system initiation and iteration completion.

Semaphore Mechanism: Semaphores for concurrent management of shared memory resources, using mutexes for critical section protection. This mechanism ensures the integrity of the FIFO queue during the send and receive operations between processes.

Shared data structures, such as message queues and synchronization variables, are strategically placed in shared SPM sections. This arrangement, devoid of software-driven caching, ensures memory consistency and simplifies communication and synchronization. The communication library, depicted in Fig. 4, efficiently handles multicore synchronization without an OS scheduler, facilitating FIFO communication in DF computing.

4 Experimental Setup

Two evaluation applications, modeled with the SDF MoC, are developed and each actor is benchmarked to compute its average execution time. To obtain the average time, we profile the actor execution time individually through a loop of a hundred iterations. Two canonical forms are generally present in a dataflow graph, and the two forms of graphs have been chosen for the application modeling. The first form is a sequence that shows pipeline parallelism. The second graph represents a fork/join topology, highlighting task and pipeline parallelism. The first form is a sequence with which we can show pipeline parallelism. The

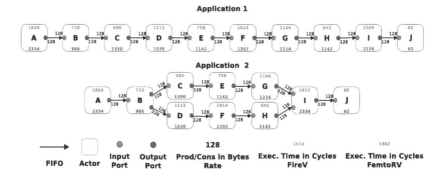

Fig. 6. The two application topologies used for experimental evaluation.

second graph presents a fork/join topology, highlighting task and pipeline parallelism. Each actor represents a distinct operation on arrays of a predefined size. Actor A initializes an array by assigning values that increase linearly, then updates the initial value based on factorial calculations. Actor B performs a subtraction operation on its second data point, while Actor C cubes its third value. Actor D reverses vector elements, and Actor E amplifies data by doubling each value. Actor F uniquely adds the data point's position to its value, and Actor G restricts values within a byte range using a modulus operation; meanwhile, Actor H and Actor J square and double specific data points in the dataset. Lastly, Actor I performs operations between two successive data in the vector.

Scratchy configurations with up to 3 cores are deployed to study whether these configurations offer nondegenerate alternatives for executing parallel workloads with varied resource/performance trade-offs. By nondegenerate alternatives, we mean that the solution on the Pareto front has a unique set of objective values. The architecture generator is configured to produce five Scratchy topologies, each embedding 1 to 3 PEs (Fig. 5). The architectural topologies are chosen to illustrate the interconnect customization property. From all scenarios, the type of PE can be selected between FemtoRV[6] and FireV[7] cores to assess the impact of core heterogeneity. The two tiny cores are open-source, resource-efficient, and integrated into the LiteX project. Both cores are clocked at 50 MHz, as is the rest of the platform. The average read and write speeds for accessing a buffer of 128 Bytes are measured at respectively 0.57 Byte/cycle and 0.59 Byte/cycle on the FireV core, and 0.42 Byte/cycle and 0.42 Byte/cycle on the FemtoRV core.

We experiment with the class of Scratchy architectures on an Intel DE10-Lite board with a MAX10 FPGA embedding 204 *kBytes* of internal BRAM memory. PEs, each consisting of a core and allocated memory for code storage and execution, is allocated a minimum of 32 *KBytes* of ROM for the bootloader and 24 *KBytes* of SRAM. The SRAM is divided into 8 *KBytes* for the bootloader data section and 16 *KBytes* for user code.

---

[6] https://github.com/BrunoLevy/learn-fpga/tree/master/FemtoRV.
[7] https://github.com/sylefeb/Silice/tree/master/projects/fire-v.

The experiments show that the FireV core can execute all types of actors faster than the FemtoRV core, except for actor J (Fig. 6). We measure the SRAM access speed to evaluate memory read and write speeds of two processors. This is done by writing a uniform data pattern for sequential memory access and calculating the speed in bytes per second by comparing the elapsed time to the memory range.

Using the PREESM multicore scheduling tool, applications are scheduled for execution on 1 to 3 cores using a list scheduling heuristic. Inter-PE communications are performed through SPMs as described in Sect. 3.2, and intra-PE communications employ temporary buffer storage. For application 1 built as a pipeline of actors, Cycles Per Graph Iteration (CPGI) is enhanced by exploiting pipelining when multiple cores are available. For Application 2, scheduling can only exploit task parallelism or mixed pipeline and task parallelism. In addition, we also pipeline the execution for a three-core Scratchy with Actors A and B on Pa, Actors I and J on Pc, and the rest on Pb. In the employed example, delays are introduced after Actor B and before Actor I to force pipelining. In these simulations, Gantt charts are displayed for two delay configurations. Computation times and communication times are considered. A metric called CPGI is introduced, which measures the cycles required for one iteration of a graph. One may note that this SDF graph iteration includes delay-induced parallelism and thus exploits both pipelining and task parallelism. In the Design Space Exploration (DSE), selecting the optimal Scratchy architecture for each application is considered a resource/latency trade-off. In other words, the optimum is generally defined by a Pareto front consisting of multiple Scratchy configurations that provide non-dominated latency/throughput performance.

5   Results and Discussion

In the results, Fi denotes a FireV core and Fe a FemtoRV core. The minimum logic resources required for implementing a one-PE system can be observed by examining the configuration resources A_1. The results of the synthesis show that the FireV processor uses more resources than the FemtoRV processor by approximately 26%, and the most resource intensive architecture has been synthesized in less than 5 min. The difference between the two cores is also reflected in the execution speed of the application actors. Analyzing the resources required for the five generated architectures reveals that the resource overhead for busses, arbiters, and memory control is very small w.r.t. PE resources, demonstrating Scratchy's lightweight communication nature. The interconnection overhead is approximately 2% for a two-core Scratchy and 20% for a three-core Scratchy. Thus, the transition from a one-core system to a two-core and three-core system results in doubling and then tripling resources with a non-negligible communication overhead. The overhead introduced by the interconnection of the cores helps determine the resources required to build a Scratchy architecture based on the configuration with one PE and the number of PE in the configuration. A much lower overhead also appears when creating custom communication with

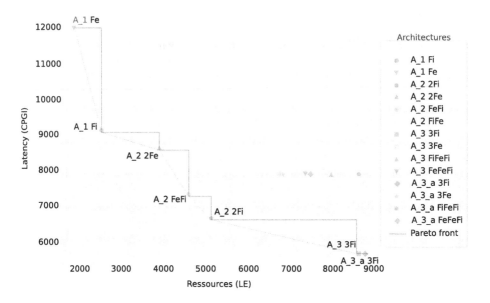

Fig. 7. CPGI vs. logic elements for each Scratchy topology on Application 1.

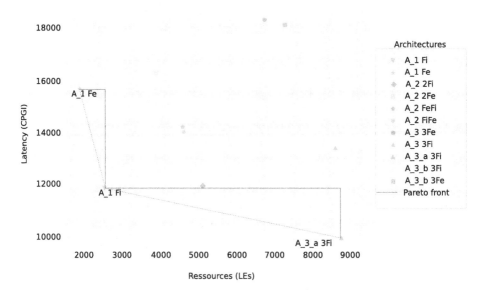

Fig. 8. CPGI vs. logic elements for each Scratchy topology on Application 2.

multiple buses. For a three-core architecture (A_3, A_3_a, A_3_b), the additional hardware cost of interconnecting the additional SPM is less than 2% for homogeneous cases (3 FireV cores or 3 FemtoRV cores) compared to A_3. This result motivates custom interconnection following the needs of the applications and system constraints.

Scratchy makes constructing advanced Design Space Exploration (DSE) processes possible, as shown in Fig. 7. Results demonstrate that the A_3_a setup, which utilizes a specialized communication approach with homogeneous FireV processors, outperforms the other configurations considering cycles per graph iteration (CPGI). The results also show that the 6 configurations are nondegenerate, appearing on the Pareto front when considering logic elements versus CPGI. The A_3 configuration closely follows the A_3_a best CPGI setup in performance. Although their execution times are identical, they differ in the time taken for communications. Total communication times across all cores decrease by tens of cycles for the FireV configuration and by nearly a hundred cycles for the FemtorV configuration. The communication time is the sum of the polling, synchronization, and copy time. Synchronization time indicates the time spent managing the synchronization elements of the FIFO semaphores.

Scratchy DSE (DSE across the design space of possible Scratchy configurations) also demonstrates considering the graph topology and the properties of the workload when choosing a platform.

The results on Application 1 show that the configurations A_2 FiFe and A_2 FeFi have similar two-core architectures but different actor mappings. One helps to shorten the heaviest workload, while the other does the opposite, shortening the lightest workload, increasing waiting times, and reducing performance. A similar analysis applies to three-core architectures, where choosing a heterogeneous set-up does not offer benefits compared to the homogeneous FemtoRV case. In summary, DSE reveals that for the SDF graph shapes of Application 1, having a heterogeneous structure does not provide benefits and is not the best solution in a Scratchy architecture. A more effective approach is to use the highest-performance cores in a homogeneous configuration. A_2 with homogeneous FireV cores provides a good compromise between resources and CPGI.

The study results on Application 2 presented in Fig. 8 show that gains from employing multiple cores are more limited. Indeed, graph scheduling does not adapt well to available resources, exploiting less parallelism than is available. A single-core system can outperform other multicore configurations. However, when graph execution is pipelined, better results are obtained. This is particularly true for homogeneous A_2 with FireV cores. As in the previous case with Application 1, we can also observe that customizing communications reduces communication times, which impacts the CPGI. Figure 8 shows the difference in CPGI between the homogeneous A_3 and A_3_b architectures. As execution times are the same, the difference is in communication times. Similarly to Application 2, we note that we have a difference in the order of hundreds of cycles for the two-core configurations ~ 200 cycles for FireV and ~ 300 cycles for FemtoRV). Times are shorter when communication is customized.

In conclusion, the experiments on the two example applications and the resource analysis show that the customizable Scratchy framework with its software support is capable of providing for a unique application a large set of potential platforms with nondegenerate performance in a CPGI versus Logic Elements (LEs) Pareto plane. The efficiency of Scratchy depends on the ability to utilize cores and, therefore, on placement/scheduling decisions, the workloads of actors, and application dependencies.

6   Conclusion

In this paper, a new class of architectures called Scratchy is introduced. Scratchy architectures are easily generated and designed explicitly for SDF-modeled applications. It allows the creation of various architectural topologies, with software support, including middleware and inter-PE communication libraries.

The paper demonstrates the features and utility of Scratchy by generating five architecture topologies with 1 to 3 PEs on an embedded Intel MAX10 FPGA, considering two representative application graph forms. The results show that Scratchy can generate a set of Pareto-optimal architectures when considering the Cycles Per Graph Iteration (CPGI) versus logic resources.

In future work, we will scale the number of cores on complex application topologies, optimize graph cuts for pipelining execution, and customize resources to improve the efficiency and sustainability of Scratchy architectures.

References

1. Amarasinghe, S., et al.: Language and compiler design for streaming applications. Int. J. Parallel Prog. (2005). https://doi.org/10.1007/s10766-005-3590-6
2. Amid, A., et al.: Chipyard: integrated design, simulation, and implementation framework for custom SoCs. IEEE Micro (2020). https://doi.org/10.1109/MM.2020.2996616
3. Asanović, K., et al.: The rocket chip generator. Technical report, EECS Department, University of California, Berkeley (2016)
4. Balkind, J., et al.: Openpiton at 5: a nexus for open and agile hardware design. IEEE Micro (2020). https://doi.org/10.1109/MM.2020.2997706
5. Balkind, J., et al.: Openpiton: an open source manycore research framework. SIGARCH Comput. Archit. News (2016). https://doi.org/10.1145/2980024.2872414
6. Banakar, R., Steinke, S., Lee, B.S., Balakrishnan, M., Marwedel, P.: Scratchpad memory: a design alternative for cache on-chip memory in embedded systems. In: Proceedings of the Tenth International Symposium on Hardware/Software Codesign. CODES 2002 (IEEE Cat. No. 02TH8627) (2002). https://doi.org/10.1145/774789.774805
7. Ghasemi, A.: Notifying memories for dataflow applications on shared-memory parallel computer. Ph.D. thesis, Université de Bretagne Sud (2022). https://tel.archives-ouvertes.fr/tel-03704297v2

8. Ghasemi, A., Cataldo, R., Diguet, J.P., Martin, K.J.M.: On cache limits for dataflow applications and related efficient memory management strategies. In: Workshop on Design and Architectures for Signal and Image Processing (14th Edition). Association for Computing Machinery (2021). https://doi.org/10.1145/3441110.3441573

9. Hennessy, J.L., Patterson, D.A.: A new golden age for computer architecture. Commun. ACM (2019). https://doi.org/10.1145/3282307

10. Kermarrec, F., Bourdeauducq, S., Lann, J.L., Badier, H.: Litex: an open-source SoC builder and library based on migen python DSL. CoRR (2020). https://api.semanticscholar.org/CorpusID:199423893

11. Krishnasamy, E., Varrette, S., Mucciardi, M.: Edge computing: an overview of framework and applications. Technical report, PRACE aisbl, Bruxelles, Belgium (2020)

12. Kurth, A., Capotondi, A., Vogel, P., Benini, L., Marongiu, A.: HERO: an open-source research platform for HW/SW exploration of heterogeneous manycore systems. In: Proceedings of the 2nd Workshop on AutotuniNg and aDaptivity AppRoaches for Energy Efficient HPC Systems. ACM (2018). https://doi.org/10.1145/3295816.3295821

13. Liu, T., Tanougast, C., Weber, S.: Toward a methodology for optimizing algorithm-architecture adequacy for implementation reconfigurable system. In: 2006 13th IEEE International Conference on Electronics, Circuits and Systems, pp. 1085–1088 (2006). https://doi.org/10.1109/ICECS.2006.379627

14. Martin, K.J.M., Rizk, M., Sepulveda, M.J., Diguet, J.P.: Notifying memories: a case-study on data-flow applications with NoC interfaces implementation. In: Proceedings of the 53rd Annual Design Automation Conference. ACM, New York (2016). https://doi.org/10.1145/2897937.2898051

15. Maurer, P.: The florida hardware design language. In: IEEE Proceedings on Southeastcon (1990). https://doi.org/10.1109/SECON.1990.117849

16. Medvidovic, N., Taylor, R.: A classification and comparison framework for software architecture description languages. IEEE Trans. Software Eng. (2000). https://doi.org/10.1109/32.825767

17. Mellor-Crummey, J.M., Scott, M.L.: Algorithms for scalable synchronization on shared-memory multiprocessors. ACM Trans. Comput. Syst. (1991)

18. Niang, P., Grandpierre, T., Akil, M., Sorel, Y.: AAA and SynDEx-Ic: a methodology and a software framework for the implementation of real-time applications onto reconfigurable circuits. In: Becker, J., Platzner, M., Vernalde, S. (eds.) FPL 2004. LNCS, vol. 3203, pp. 1119–1123. Springer, Heidelberg (2004). https://doi.org/10.1007/978-3-540-30117-2_143

19. Pelcat, M., Desnos, K., Heulot, J., Guy, C., Nezan, J.F., Aridhi, S.: Preesm: a dataflow-based rapid prototyping framework for simplifying multicore DSP programming. In: 2014 6th European Embedded Design in Education and Research Conference (EDERC) (2014). https://doi.org/10.1109/EDERC.2014.6924354

20. Petrisko, D., et al.: BlackParrot: an agile open-source RISC-V multicore for accelerator SoCs. IEEE Micro (2020). https://doi.org/10.1109/MM.2020.2996145

21. Rouxel, B., Skalistis, S., Derrien, S., Puaut, I.: Hiding communication delays in contention-free execution for SPM-based multi-core architectures. In: Euromicro Conference on Real-Time Systems (2019). https://doi.org/10.4230/LIPIcs.ECRTS.2019.25

22. Zuckerman, J., Mantovani, P., Giri, D., Carloni, L.P.: Enabling heterogeneous, multicore SoC research with RISC-V and ESP. arXiv (2022). https://doi.org/10.48550/arXiv.2206.01901

Digital Signal Processing Design
for Reconfigurable Systems

Standalone Nested Loop Acceleration on CGRAs for Signal Processing Applications

Chilankamol Sunny[1] , Satyajit Das[1(✉)] , Kevin J. M. Martin[2] ,
and Philippe Coussy[2]

[1] IIT Palakkad, Kerala, India
`112004004@smail.iitpkd.ac.in`, `satyajitdas@iitpkd.ac.in`
[2] Univ. Bretagne-Sud, UMR 6285, Lab-STICC, 56100 Lorient, France
{`kevin.martin,philippe.coussy`}`@univ-ubs.fr`

Abstract. Coarse-Grained Reconfigurable Array (CGRA) architectures
are becoming increasingly popular as low-power accelerators in compute
and data intensive application domains such as security, multimedia, signal processing, and machine learning. The efficiency of a CGRA is determined by its architectural features and the compiler's ability to exploit
the spatio-temporal configuration. Numerous design optimizations and
mapping techniques have been introduced in this direction. However, the
execution model has been overlooked, despite its critical role in ensuring the efficient acceleration of applications. Most of the existing CGRA
implementations follow a hosted approach i.e., they execute the modulo
scheduled innermost loop, entrusting outer loops to the host processor.
This increases synchronization overhead with the host, mitigating the
benefits of acceleration provided by the CGRA. In this paper, we propose a compilation flow that supports efficient standalone execution of
nested loops. Experiments show that the standalone execution model
leads to a maximum of $12.33\times$ and an average of $6.75\times$ performance
improvement compared to the existing hosted execution model. In the
proposed model, energy consumption is reduced up to $14.49\times$ compared
to that of the hosted one. We also compared our results with state-of-
the-art standalone execution that uses loop flattening and achieved a
maximum of $4.80\times$ speed up with an average of $2.80\times$.

Keywords: Coarse grained reconfigurable array (CGRA) · Nested
loop acceleration · Standalone execution model

1 Introduction

Due to the architectural elasticity, Coarse-Grained Reconfigurable Array
(CGRA) architectures offer high performance, energy efficiency, and flexibility [13]. A typical CGRA integrated system consists of an array of interconnected
processing elements (PEs) tightly/loosely coupled with a host CPU, a context,

© The Author(s), under exclusive license to Springer Nature Switzerland AG 2024
T. Dias and P. Busia (Eds.): DASIP 2024, LNCS 14622, pp. 83–95, 2024.
https://doi.org/10.1007/978-3-031-62874-0_7

and a data memory. Each PE is composed of a word-level configurable functional unit (FU), a regular register file, an instruction memory, and routers at the input and output. CGRAs have been proposed to cater to the needs of both High Performance Computing (HPC) and Low Power Computing (LPC) domains with various architectural and compilation novelties. Architectural improvements like dynamic voltage and frequency scaling (DVFS) [19], approximate arithmetic units [1] and efficient memory hierarchies [2] have been proposed to improve the performance and energy efficiency of CGRAs. To improve the performance of the compute-intensive applications, several loop optimization techniques have been adopted in the compilation flow, such as loop unrolling [6], modulo scheduling [8] and polyhedral loop optimizations [12]. However, most of the state-of-the-art CGRA compilation flow [8,9,18] uses modulo scheduling loop optimization due to its good performance for the innermost loop.

The majority of the existing CGRAs focus on the optimized innermost loop execution, entrusting outer loops to the host processor. This increases the synchronization overhead, diminishing the benefits of acceleration provided by the CGRA [4]. Hence, optimized mapping techniques and improved architectural designs are not sufficient to guarantee the best performance. The execution model, which defines how tasks or processes are partitioned, scheduled, and executed (includes configuration and execution time) on CGRAs, is equally important. This paper discusses and analyses the standalone and hosted execution models for CGRAs. In the standalone model, the entire nested loop structure is run on the CGRA with no host intervention. In the hosted model, CGRAs execute only the innermost loop letting the host processor execute the outer loops.

The major contributions of this work are: (a) An explorative study on the impact of execution models in determining the performance and energy efficiency of CGRAs. We demonstrate that a highly efficient mapping technique may not be sufficient to guarantee the best performance. The execution model is equally important, especially for kernels with deeply nested loops which is the case with most modern signal processing and AI applications [20]. (b) A compilation flow for CGRAs that supports the standalone execution of nested loops. The proposed approach modulo schedules the innermost loop using traditional modulo scheduling algorithms and executes loops at all levels of loop nesting on the CGRA, with no host intervention.

The rest of the paper is organized as follows. Section 2 presents the background and related works. Section 3 introduces the proposed compilation flow. The experiments and results are discussed in Sect. 4. Section 5 concludes the paper.

2   Background and Motivation

CGRAs, due to their architectural specialization, efficiently execute the pipelined innermost loop (single loop). Traditional CGRAs target to accelerate only the innermost loop for applications with nested loops, leaving outer loops for the

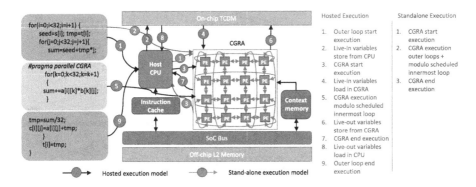

Fig. 1. Hosted and standalone execution model for CGRA loosely coupled with CPU

host processor. This is referred to as *hosted* execution model (Fig. 1) in this paper. In this execution model, the variables needed for the CGRA to execute the innermost loop (*live-in variables*), and the variables processor needs from CGRA (*live-out variables*) are transferred through shared memory. The overhead due to added memory operations and communication for synchronization are shown in Fig. 1. To minimize the overhead, proposals like [10,12,21] perform several loop transformations (i.e. polyhedral transformation, loop flattening, loop fission). However, with the growing complexity of the loop nests in signal processing applications, in addition to the transformations, we need mechanisms to minimize the host intervention. Hence, the *standalone* execution of the entire loop nests is the ideal solution as presented in Fig. 1.

For the hosted execution of CGRAs, the works [5,8,14] perform *modulo scheduling* on the innermost loop. It is a software pipelining technique that facilitates overlapped execution of different iterations of a loop. The goal of modulo scheduling is to find a schedule of operations from different iterations of the innermost loop that can be repeated in a short interval called initiation interval (II), expressed in cycles. The data flow graph (DFG) formed by this repeating schedule is referred to as Modulo Data Flow Graph (MDFG). The set of operations that are executed once before and after the MDFG, form a couple of DFGs and are referred to as prologue and epilogue respectively [17]. As the hosted execution only executes the innermost loop, mapping of the MDFG onto the target CGRA is considered a DFG mapping problem [8,9]. The prologue and epilogue mappings are adapted from the mapped MDFG [16]. Figure 2(d) shows the replication of MDFG mapping in Fig. 2(c). The innermost loop DFG in this example is presented in Fig. 2(a). Figure 2(b) shows the *prologue epilogue*, and *MDFG* after modulo scheduling. In this example, the two-cycle-long MDFG mapping (Fig. 2(a)) is replicated twice to prepare the prologue mapping, resulting in a schedule length of 4 (Fig. 2(d)). Similarly, the epilogue mapping is also prepared from the MDFG mapping with a schedule length of 2. However, replicating the MDFG mapping does not always guarantee the optimum solution. As shown in Fig. 2(e), a mapping solution of schedule length 3 is obtained by mapping the

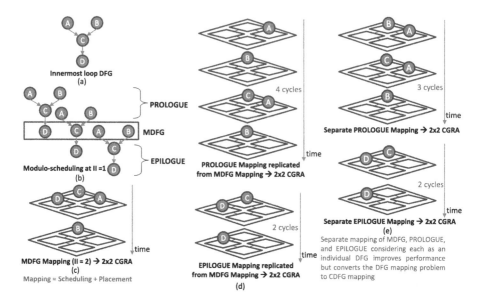

Fig. 2. Modulo scheduling example with MDFG, prologue, epilogue mapping

prologue DFG directly onto the CGRA rather than adapting from the MDFG mapping which resulted in a higher schedule length. The larger schedule length may seem very less in a single-nested loop. However, for loop nests if the iteration count increases, the cumulative effect of the larger schedule results in degraded performance. Thankfully, prologue and epilogue DFGs always contain a smaller number of operations per cycle compared to the MDFG due to the inherent construct of MDFG. As the number of nodes is less, mapping the prologue and epilogue as individual DFGs results in a lower schedule length. This is a CDFG mapping problem where, the prologue, epilogue, and MDFG are considered as individual DFGs, and while mapping, control flow between the DFGs needs to be satisfied. In this paper, the standalone execution is achieved by the mapping of the entire application CDFG along with the modulo scheduled innermost loop kernel.

The state-of-the-art solution, Cheng et al [18] proposes support for the standalone execution where the loop nests are flattened into a single-nested loop (Fig. 3) to facilitate the DFG mapping. The resultant DFG is modulo scheduled and executed on the CGRA. The solution suffers from inflated DFG when the number of loops gets increased causing increased II and high energy consumption. IPA [3] approach proposes to support the standalone execution of nested loops using *register allocation-based direct-mapping* of CDFG onto CGRAs achieving good performance and energy results. However, the compilation flow proposed in that work uses partial loop unrolling instead of modulo scheduling the innermost loop. In this paper, we extend the solution proposed

```
Loop1: for (I=0; i<M ; i++){
          sum=0;
Loop2: for (j=0; j<N ; j++)
          sum += array_in[i][j];
array_out[i] = sum;
}
```

Nested loop in proposed approach

The modulo scheduled loop bodies in
both the approaches are highlighted.

```
Loop1: for (n=0; n < M * N ; n++){
          i = n / N ;
          j = n % N ;
          if (j==0)
                    sum=0;
          sum += array_in[i][j];
          if (j==N -1)
                    array_out[i] = sum;
}
```

Flattened loop used in [18]

Fig. 3. Example of where modulo scheduling is applied in the proposed and loop transformation based standalone execution models

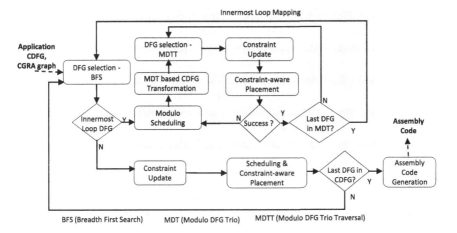

Fig. 4. Proposed compilation flow supporting standalone execution of nested loops and modulo scheduling of innermost loop

in the IPA [3] to support the standalone execution of the entire loop nests with the innermost loop modulo scheduled (Fig. 3).

3   Proposed Approach

3.1   Overview

We introduce a novel compilation flow for CGRAs that supports standalone execution of the entire application containing loop nests. The innermost loop uses modulo scheduling for the pipelined implementation.

As the problem is defined as a CDFG mapping problem, we adapt the register allocation-based approach proposed in Das et al [3] where the basic blocks (BB) (individual DFGs) are mapped onto CGRA with a simultaneous scheduling and placement algorithm. The variables that are used in multiple BBs (*symbol variables*) are mapped in registers using *target location*, and *reserved location* constraints in placement. In this paper, we propose a mapping of modulo scheduled innermost loop DFG along with the other BBs in the CDFG. The primary challenge in this strategy is to integrate the MDFG, prologue, and epilogue generated by modulo scheduling in the original kernel CDFG and traverse efficiently to find valid mappings for the entire CDFG. To meet this challenge, we have designed a CDFG transformation and traversal technique which are explained in the following section.

The proposed compilation flow comprises two tracks as depicted in Fig. 4, one for the innermost loop mapping and the other for mapping the rest of the CDFG. The first step in the compilation flow is to choose the BB for mapping. The selection is done by the breadth-first search (BFS) traversal of the CDFG, the technique proven to generate the least number of constraints in CDFG mapping [3]. If the selected BB (DFG) corresponds to the innermost loop, it is first modulo scheduled and then placed onto the CGRA, respecting the register allocation constraints imposed by the mappings of already mapped BBs. Every other BB is mapped by following a simultaneous scheduling and placement approach [3]. The data integrity over different BB mappings is maintained by the constraint-aware placement technique. The constraint update step in the compilation flow sets the register allocation constraints that guide the choice of registers in the placement process. These constraints ensure reserved usage of registers for variables that are used in multiple BBs. Once all BBs are mapped, the compiler generates the assembly code for the entire CDFG mapping.

Simultaneous Scheduling and Placement. A priority-based list scheduling algorithm is used to schedule the DFG nodes and an incremental version of Levi's algorithm [11] is used for placement and routing (binding). Failing to bind a node, the compilation flow transforms the DFG dynamically and continues with the mapping process. If a transformation that improves the mapping possibilities cannot be identified, backtracking is performed, and mapping restarts with a new mapping context. This is done by choosing the next BB from the set of previously mapped BBs. If the mapping is successful, a stochastic pruning is applied on the partial mapping set to prevent it from growing exponentially.

3.2   Innermost Loop Mapping

If the DFG selected for mapping corresponds to the innermost loop, it is modulo scheduled [16] and placed onto the CGRA, following the innermost loop track in the compilation flow.

Modulo Scheduling. Modulo scheduling starts with computing the minimum possible II (MII), determined by the resource and recurrence constraints of the input DFG and the target CGRA. The nodes in the DFG are then modulo scheduled [16] with an II equal to MII. Failing to find a schedule or placement solution for this II, the process is restarted with an incremented II and repeated until a valid mapping is found. Modulo Scheduling a DFG splits it into three

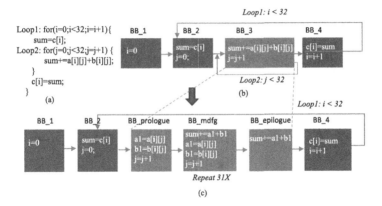

Fig. 5. MDT based CDFG transformation: (a) Sample for loop; (b) corresponding CDFG; (c) CDFG after MDT based transformation

DFGs, the prologue, MDFG, and epilogue, together called Modulo-DFG-Trio (MDT) in the discussion hereafter. This calls for a local graph transformation in the CDFG.

MDT Based CDFG Transformation. Fig 5 illustrates the MDT based CDFG transformation we introduce to apply modulo scheduling in conjunction with direct CDFG mapping. The figure presents a sample *for* loop and the corresponding CDFG. Fig. 5 (c) gives the transformed CDFG in which the innermost loop DFG is replaced with the MDT. Edges connecting these DFG nodes are set such that the control flows from the outer loop BB to prologue, from prologue to MDFG, and from MDFG to epilogue. The epilogue DFG is connected to the immediate successor of the innermost loop DFG in the original CDFG. MDFG is executed multiple times, determined by the number of times the innermost loop DFG is unrolled to prepare the modulo schedule.

DFG Selection by Modulo DFG Trio Traversal (MDTT). Unlike the conventional approach of preparing the prologue and epilogue mappings from the MDFG mapping, we propose to map the three DFGs separately (respecting the modulo schedule generated in the previous step). As shown in the motivating example (Fig. 2(e)), mapping the DFGs separately helps in achieving shorter schedule lengths for the prologue and/or epilogue, leading to improved performance and energy efficiency. This is now a nested CDFG mapping problem, solved by employing the register allocation approach. However, in the case of the CDFG formed by MDT, all variables in all three DFGs are symbol variables (variables used in multiple BBs), by the inherent construct of modulo schedule. One location constraint is generated each time a symbol variable is placed for the first time. Hence, the ordering of DFGs for mapping is crucial (especially in the case of MDT) to ensure the flexibility of mapping. We propose MDT

Table 1. Modulo DFG Trio (MDT) Traversal explained

Let P, M and E be the set of all variables in prologue, MDFG and epilogue respectively; S be the set of all symbol variables (variables used in multiple BBs) in the MDT.	
$M = P \cup E$ $S = (P \cap M) \cup (M \cap E) \cup (P \cap E)$ $P \cap M = P$; $M \cap E = E$; $P \cap E \subseteq M$ Therefore $S = P \cup E \cup (P \cap E) = M$	#symbol variables in prologue = $n(S \cap P) = n(M \cap P) = n(P)$ in MDFG = $n(S \cap M) = n(M \cap M) = n(M)$ in epilogue = $n(S \cap E) = n(M \cap E) = n(E)$ $n(M) = n(P) + n(E) - n(P \cap E) =>$ $n(M) > n(P)$ & $n(M) > n(E)$
Modulo DFG Trio Traversal (MDTT) MDFG ->Prologue ->Epilogue; if $n(P) > n(E)$ MDFG ->Epilogue ->Prologue; otherwise	

traversal (MDTT) technique, explained in Table 1, for DFG selection from the MDT. Mapping BBs with a greater number of symbol variables earlier helps to reduce the number of location constraints [3]. Hence, MDTT chooses the MDFG first, the DFG with the highest number of symbol variables in the MDT. Next, it selects the prologue, if the number of variables in the prologue is more than that of the epilogue and epilogue otherwise.

Constraint Update and Placement. While mapping the MDFG, the compiler fetches the register allocation constraints generated by the previously mapped BBs and finds a placement solution that meets these constraints. The next DFG in the MDT is placed considering the constraints generated by the MDFG as well as the previous BBs. Similarly, the mapping of the next BB in the outer loop will be bound by the constraints set by the MDT as well. This technique of constraint-aware placement maintains the data integrity between the separately mapped BBs of the CDFG.

4   Experiments and Results

In this section, we evaluate the performance of the standalone execution vs the hosted execution. In the hosted execution, the innermost loop is modulo scheduled using *EpiMap* [8] and executed onto CGRA. The outer loops of the applications are run and controlled by the host CPU. The proposed standalone execution runs the entire application onto CGRA without any interruption from the CPU. To be fair with the comparison, the modulo scheduled innermost loop is mapped using *EpiMap* like approach. Here, the prologue and epilogue DFGs are mapped separately as individual DFGs instead of adapting from the MDFG mapping. We also present a performance comparison study with a state-of-the-art standalone solution by using loop transformation modeled in Cheng et al [18], which reduces the communication and memory overhead by flattening loop nests into a single-nested loop.

Table 2. Listed kernels and their loop characteristics

Kernel	Nest depth	Max # Iterations	Loop nest structure
Matrix Multiplication	3	32X32X32 = 32 768	Imperfect
Histogram Equalization	2	80X60 = 4 800	Perfect
2D Non-Separable Filter	4	58X78X3X3 = 40 716	Imperfect
FIR Filter	2	190X10 = 1 900	Imperfect
DCT	3	8X8X8 = 512	Imperfect
Bicg	2	32X32 = 1 024	Imperfect
2D Convolution	4	58X38X3X3 = 19 836	Imperfect
Sobel	4	62X62X3X3 = 34 596	Imperfect

4.1 Experimental Setup

The proposed compilation flow for the standalone execution is implemented by using Java and Eclipse Modeling Framework (EMF). The target CGRA for all our experiments is a 4×4 PE array configuration of state-of-the-art Integrated Programmable Array (IPA) architecture [4], loosely coupled with a host CPU as shown in Fig. 1. The CPU is a RISCV [7] core based on a four pipeline stages micro-architecture optimized for energy- efficient operations in digital signal processing (DSP). A multi-banked tightly coupled data memory (TCDM) facilitates data communication between IPA and the CPU. Energy results are computed using the switching activity obtained by simulating the placement-and-routed netlist design. The CGRA design is synthesized with Cadence Genus Synthesis Solution using 90nm CMOS technology libraries. Placement and Routing are performed using Cadence Innovus and power analysis is done with Cadence Voltus at the supply of 0.9 V, in typical process conditions. A set of loop-intensive signal processing kernels including those from PolyBench [15] benchmark suite is chosen for our experiments. Table 2 features these kernels with the number of levels of loop nesting (depth of nesting), maximum number of iterations, and loop nest structures present. Nesting of loops is perfect if all the assignment statements are in the innermost loop otherwise it is imperfect nesting. Due to the additional assignments in the outer loops, the imperfect nested loops usually have more live-in and live-out variables.

4.2 Results and Discussion

Performance Comparison of Different Execution Models. Table 3 presents execution latency in cycles on hosted and standalone execution models. The standalone execution achieves an average speed-up of 6.75× with a maximum of 12.33× over the hosted model. As discussed in the previous sections, hosted

Table 3. Performance and Energy comparison between hosted and standalone execution of applications with innermost loop modulo scheduled

Kernel	Execution Model	Latency (cycles)	Throughput (Mbps)	Energy (μJoule)	Speed-up	Throughput gain	Energy gain
Matrix	Hosted	464 159	1.24	287.64	4.10x	4.10x	4.88x
Multiplication	Standalone	113 310	5.07	58.97			
Histogram	Hosted	25 225	106.83	15.31	1.63x	1.63x	1.90x
Equalization	Standalone	15 484	174.03	8.06			
2D Non-Separable	Hosted	2 783 615	0.97	1702.74	12.33x	12.33x	14.49x
Filter	Standalone	225 768	11.94	117.49			
FIR Filter	Hosted	43 365	2.59	26.37	6.87x	6.87x	8.03x
	Standalone	6,308	17.80	3.28			
DCT	Hosted	14 450	2.49	8.79	5.14x	5.14x	6.01x
	Standalone	2 813	12.77	1.46			
Bicg	Hosted	12 452	2.89	7.56	1.93x	1.93x	2.25x
	Standalone	6 451	5.57	3.36			
2D Convolution	Hosted	1 352 406	1.00	845.47	10.70x	10.70x	12.85x
	Standalone	126 446	10.66	65.80			
Sobel	Hosted	2 534 676	0.91	1583.88	11.32x	11.32x	13.60x
	Standalone	2 23 844	10.27	116.49			

execution incurs a communication, and memory exchange overhead with the host CPU, resulting in increased latency. The overhead is directly related to the total number of control transfers and memory operations performed which in turn depends on the kernel structure like the outer loop count and the number of live-in and live-out variables. This is evident from the speed-up figures that standalone execution reports. For instance, the highest speed-ups are achieved on 2D Non-Separable Filter and 2D convolution kernels that feature the highest number of outer loop iterations and live-in/live-out variables among the kernels we considered. The average compilation times for the hosted and standalone executions are 11.73 and 18.61 seconds respectively.

Energy Consumption Comparison of Different Execution Models. Table 3 lists the energy results for the hosted and standalone execution models. The results demonstrate that the memory operations performed in the live-in and live-out phases of the hosted execution significantly increase energy consumption. The standalone model achieves an average of 8.00× reduction in energy consumption over hosted execution by eliminating the communication overhead with the host CPU. A maximum reduction of 14.49× is reported for the 2D non-separable filter as it performs the highest number of memory operations to transfer the live-in and live-out variables between the host and the CGRA.

Performance Comparison with Loop Transformed Standalone Execution. Fig. 6(a) compares the execution latencies between two standalone approaches. Both approaches use modulo scheduled innermost loop. However, Cheng et al [18] transforms the loop nests into a single-nested loop by flattening whereas the proposed approach maps the CDFG directly using constraint aware placement. The results show that the proposed approach achieves an average of 2.80× (with a maximum of 4.80×) speed up over the loop transformation approach. The innermost loop DFG sizes of different kernels that are modulo scheduled in the two approaches are presented in Fig. 6(b). Due to inflated DFGs, transformation-based approaches deal with larger DFGs to be mapped, hence the performance deteriorates.

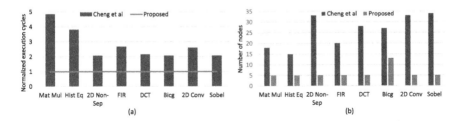

Fig. 6. Comparison between two standalone approaches (loop flattening-based vs proposed) (a) Performance comparison; (b) Comparison between the Number of nodes in the DFG to be modulo scheduled

5   Conclusion

This paper presented a standalone execution model for the acceleration of nested loops on CGRAs. The main contribution is the compilation flow that: i) modulo-schedules the innermost loop and achieves reduced execution latencies by separately mapping prologue, MDFG, and epilogue DFGs. ii) combines modulo-scheduling with direct CDFG mapping for efficient standalone execution of nested loops. The results show that the standalone model leads to a maximum of 12.33× and an average of 6.75× performance improvement compared to the hosted model. The proposed approach reports also a maximum of 4.80× and an average of 2.80× speed-up when compared to the standalone solution.

References

1. Akbari, O., Kamal, M., Afzali-Kusha, A., Pedram, M., Shafique, M.: X-cgra: An energy-efficient approximate coarse-grained reconfigurable architecture. IEEE Trans. Comput.-Aided Design Integrated Circ. Syst. **39**(10) (2019). https://doi.org/10.1109/TCAD.2019.2937738
2. Dai, L., Wang, Y., Liu, C., Li, F., Li, H., Li, X.: Reexamining cgra memory subsystem for higher memory utilization and performance. In: 2022 IEEE 40th International Conference on Computer Design (ICCD). IEEE (2022).https://doi.org/10.1109/ICCD56317.2022.00017
3. Das, S., Martin, K.J., Coussy, P., Rossi, D., Benini, L.: Efficient mapping of cdfg onto coarse-grained reconfigurable array architectures. In: 2017 22nd Asia and South Pacific Design Automation Conference (ASP-DAC). IEEE (2017).https://doi.org/10.1109/ASPDAC.2017.7858308

4. Das, S., Martin, K.J., Rossi, D., Coussy, P., Benini, L.: An energy-efficient integrated programmable array accelerator and compilation flow for near-sensor ultralow power processing. IEEE Trans. Comput.-Aided Design Integrated Circ. Syst. **38**(6) (2018).https://doi.org/10.1109/TCAD.2018.2834397

5. Dave, S., Balasubramanian, M., Shrivastava, A.: Ramp: resource-aware mapping for cgras. In: Proceedings of the 55th Annual Design Automation Conference (2018).https://doi.org/10.1145/3195970.3196101

6. Dragomir, O.S., Stefanov, T., Bertels, K.: Loop unrolling and shifting for reconfigurable architectures. In: 2008 International Conference on Field Programmable Logic and Applications. IEEE (2008).https://doi.org/10.1109/FPL.2008.4629926

7. Gautschi, M., et al.: Near-threshold risc-v core with dsp extensions for scalable iot endpoint devices. IEEE Trans. Very Large Scale Integration (VLSI) Syst. **25**(10) (2017).https://doi.org/10.1109/TVLSI.2017.2654506

8. Hamzeh, M., Shrivastava, A., Vrudhula, S.: Epimap: using epimorphism to map applications on cgras. In: Proceedings of the 49th Annual Design Automation Conference (2012).https://doi.org/10.1145/2228360.2228600

9. Hamzeh, M., Shrivastava, A., Vrudhula, S.: Regimap: register-aware application mapping on coarse-grained reconfigurable architectures (cgras). In: Proceedings of the 50th Annual Design Automation Conference (2013).https://doi.org/10.1145/2463209.2488756

10. Lee, J., Seo, S., Lee, H., Sim, H.U.: Flattening-based mapping of imperfect loop nests for cgras. In: Proceedings of the 2014 International Conference on Hardware/Software Codesign and System Synthesis (2014).https://doi.org/10.1145/2656075.2656085

11. Levi, G.: A note on the derivation of maximal common subgraphs of two directed or undirected graphs. Calcolo **9**(4) (1973).https://doi.org/10.1007/BF02575586

12. Liu, D., Yin, S., Liu, L., Wei, S.: Polyhedral model based mapping optimization of loop nests for cgras. In: Proceedings of the 50th Annual Design Automation Conference (2013).https://doi.org/10.1145/2463209.2488757

13. Liu, L., et al.: A survey of coarse-grained reconfigurable architecture and design: taxonomy, challenges, and applications. ACM Comput. Surv. **52**(6) (2019).https://doi.org/10.1145/3357375

14. Park, H., Fan, K., Mahlke, S.A., Oh, T., Kim, H., Kim, H.s.: Edge-centric modulo scheduling for coarse-grained reconfigurable architectures. In: Proceedings of the 17th international conference on Parallel architectures and compilation techniques (2008).https://doi.org/10.1145/1454115.1454140

15. Pouchet, L.N., Grauer-Gray, S.: Polybench: the polyhedral benchmark suite (2012). http://www-roc.inria.fr/~pouchet/software/polybench

16. Rau, B.R.: Iterative modulo scheduling: An algorithm for software pipelining loops. In: Proceedings of the 27th annual international symposium on Microarchitecture (1994).https://doi.org/10.1145/192724.192731

17. Rau, B.R., Schlansker, M.S., Tirumalai, P.P.: Code generation schema for modulo scheduled loops. SIGMICRO Newsl. **23**(1-2), 158-169 (1992).https://doi.org/10.1145/144965.145795

18. Tan, C., Xie, C., Li, A., Barker, K.J., Tumeo, A.: Opencgra: an open-source unified framework for modeling, testing, and evaluating cgras. In: 2020 IEEE 38th International Conference on Computer Design (ICCD). IEEE (2020).https://doi.org/10.1109/ICCD50377.2020.00070

19. Torng, C., Pan, P., Ou, Y., Tan, C., Batten, C.: Ultra-elastic cgras for irregular loop specialization. In: 2021 IEEE International Symposium on High-Performance Computer Architecture (HPCA). IEEE (2021).https://doi.org/10.1109/HPCA51647.2021.00042
20. Wijerathne, D., Li, Z., Mitra, T.: Accelerating edge ai with morpher: an integrated design, compilation and simulation framework for cgras (2023).https://doi.org/10.48550/arXiv.2309.06127
21. Wijerathne, D., Li, Z., Pathania, A., Mitra, T., Thiele, L.: Himap: Fast and scalable high-quality mapping on cgra via hierarchical abstraction. IEEE Trans. Computer-Aided Design of Integrated Circuits and Systems **41**(10) (2021). https://doi.org/10.1109/TCAD.2021.3132551

Improving the Energy Efficiency of CNN Inference on FPGA Using Partial Reconfiguration

Zhuoer Li$^{(\boxtimes)}$ and Sébastien Bilavarn

LEAT, Université Côte d'Azur, CNRS, Sophia Antipolis, France
{zhuoer.li,sebastien.bilavarn}@univ-cotedazur.fr

Abstract. With the increasing demand for edge AI application scenarios, as the most popular deep learning models, Convolutional Neural Networks (CNNs) need advanced solutions for the deployment of highly energy-efficient implementations. This paper presents a novel approach to improve the efficiency of CNN inference on Field-Programmable Gate Arrays (FPGAs) using Partial Reconfiguration (PR). Our method deconstructs CNN topology into different layers for runtime reconfiguration with fewer resources, aiming to significantly reduce static power and overall energy consumption. To identify the conditions for practical PR efficiency, we present a thorough design space exploration study with three CNN benchmarks, each evaluated across three different implementations. The comparison results demonstrate that our PR approach can achieve up to 3.88 and 1.67 times energy savings compared to software and static hardware implementations, respectively. These results also show that the benefits of PR improve with the depth of the network, suggesting very promising levels of gains as the network gets larger and under the key conditions of using fast optimized reconfiguration controllers and methodical system-level exploration of the increased hardware implementation complexity.

Keywords: CNN accelerator · Dynamic Partial Reconfiguration · Energy Efficiency · High-level Synthesis

1 Introduction

In recent years, the popularity of Artificial Intelligence (AI) in various domains has led to a growing demand for research in power efficiency, particularly due to significant computing requirements in edge AI. Embedded deep learning models, especially Convolutional Neural Networks (CNNs), have received a lot of interest from embedded development research which quickly started to investigate reconfigurable accelerators and FPGAs to combine high performance, low energy, and flexibility. In this regard, Partial Reconfiguration (PR) is now a well-identified technique that can be also used to push FPGA adaptivity one step further. This technique allows modifying an FPGA configuration by loading another configuration file, often referred to as a "partial" BIT file, during runtime. In general,

© The Author(s), under exclusive license to Springer Nature Switzerland AG 2024
T. Dias and P. Busia (Eds.): DASIP 2024, LNCS 14622, pp. 96–109, 2024.
https://doi.org/10.1007/978-3-031-62874-0_8

the FPGA is partitioned into different areas called Reconfigurable Regions (RR) which are physically independent and could also theoretically be run in parallel. While a RR is running, it can be locally configured independently with a new partial BIT file without compromising the integrity of the other running regions. Therefore, this technique of local dynamic reconfiguration enables the configuration of more system functions on a smaller surface area, consuming less power, and allows running regions in parallel, thereby enhancing performance. Correct exploitation of this means potentially bringing significant additional benefits on top of those resulting from pure static FPGA allocation, but the counterpart is design complexity to identify relevant deployments which can quickly explode with the application complexity and the number of hardware functions.

In this work, we address this with the help of a previously defined methodology for efficient mapping of large and complex application graphs on manycore reconfigurable accelerated systems allowing dynamic and partial reconfiguration [11]. Section 2 first reviews existing works on CNN hardware acceleration in relation to dynamic reconfiguration. Section 3 addresses the global hardware software investigation methodology including the central Design Space Exploration (DSE) approach dealing with PR. Section 4 reports experiment studies of three CNN benchmarks and a comparison with other works. Section 5 finally draws the main conclusions of the results with future direction for research.

2 State of the Art

Despite the very abundant literature investigating the question of FPGA acceleration for Deep Learning, PR only started to be addressed in quite recent works. Relatively few papers have therefore investigated how to benefit from the capabilities of PR to support more specifically the computation cost of CNN processing. Ideally, PR would rely on sequentially scheduling each layer of a CNN onto one or a few RR(s) to reduce the overall size of the reconfigurable area. In other words, the entire share of hardware functions would be mapped dynamically to RRs and no static part would remain in the design, a concept we'll refer to as "reductive PR". However, in practice, mainly due to excessive reconfiguration times, a majority of works restrict the use of PR to a smaller part of the system, referred to as "functional PR".

This is for example the case in [1] (2019) where PR is used to add layers at the convolution level when needed, to allow switching dynamically from shallow to deeper topologies, providing this way an adaptive capacity to adjust the model structure and overall classification accuracy at run-time. Another technique is described in [4] (2021) where a macroblock-based PR implementation is defined to let the dynamic reconfiguration of parts of the CNN. Application to a LeNet topology addresses the reconfiguration of two different fully connected layers, enabling a switch between two network versions for either number recognition or letter recognition. Functional PR is also involved in [5] (2014) where a PR strategy is used to implement sigmoid and hyperbolic tangent functions in a Multi-Layer Perception (MLP), or in [2] (2022) and [3] (2020) to exploit different quantization schemes or bit widths (12, 10, 8, 7, 6, 5 bits).

In the category of works addressing a reductive PR approach, fpgaConvNet [6] (2016) is the first proposed approach in the literature, up to our knowledge. In their work, the authors define an automated design flow for mapping CNNs on FPGAs considering design space exploration and High-Level Synthesis (HLS). They employ a formalization technique to automatically explore parallelism such as intra-layer unrolled execution and inter-layer pipeline to tune the performance-resource tradeoff and explore the design space. It should be noted here that, in this approach, the authors intend to map each layer on the entire FPGA without partitioning them into RRs. As a result, they need to employ full FPGA reconfiguration before the execution of each layer. They address the associated reconfiguration latency problem by pipelining execution and reconfiguration to overlap and amortize the associated latency overheads. However, this solution does not *practically* remove the important reconfiguration times of the full FPGA and especially their associated energy cost. For instance, the reconfiguration latency for the entire XC7Z020 device using PCAP on a ZedBoard platform is around 28 ms. Nevertheless, the results show very promising savings and efficiency from FPGA mapping with this dynamic reconfiguration process.

[7] (2018) is a second significant contribution addressing PR. Starting out from an initial statement on excessive reconfiguration times, which range from 100 to 150 ms in their application study of an eight-layer CNN for facial landmark detection, the authors describe a DSE process in which they consider overlapping reconfiguration of parts of the inference path with other parts that are executed in software. Pooling and activation function layers run therefore in software while they alternate software and hardware executions of convolutional layers. The approach allows reducing the overall execution time to a factor of 2.24 compared to pure software execution. Finally, among this class of PR, [7] is the sole work, up to our knowledge, that considers the critical partitioning problem inherent with PR. A formal DSE approach is defined to both compute the optimal configuration of each layer and decide how to split the network into chunks that will be dynamically programmed via PR. Application on Binary Neural Networks (BNNs), specifically the CNVW1A1 Binary Neuralnet from Xilinx, reports a clear potential for resource reduction against static FPGA allocation, but with highly varying degrees of performance - occupation tradeoffs and unclear performance penalty from large reconfiguration times.

In conclusion, reductive PR is the most interesting approach for reducing FPGA logic requirements and thus extending the efficiency of static FPGA acceleration. The potential is vast for CNNs because most part of their processing can be moved to hardware, and this grows very quickly with the size of the topology. However, two important conditions must be met: i) a methodical DSE approach allowing to the identification of suitable solutions in the very large and complex design space involved with PR and ii) important optimization of reconfiguration processes are also mandatory. In the following, we investigate this by using i) a previously defined methodology for the efficient mapping on many-core reconfigurable accelerated systems allowing PR [11] and ii) a very fast optimized reconfiguration controller [10].

3   Partial Reconfiguration Flow

In this study, we investigate the potential of using PR to improve the energy efficiency of CNNs on FPGAs (Fig. 1). A SystemC simulation methodology for modeling, scheduling, and estimation of multiprocessor reconfigurable systems called FoRTReSS [11] is used to explore the mapping of CNNs on reconfigurable platforms supporting PR. In this approach, applications are represented as Control Data Flow Graphs (CDFGs) consisting of several tasks, each with multiple possible implementations that can be hardware-mapped to the reconfigurable area or software-mapped to processor cores.

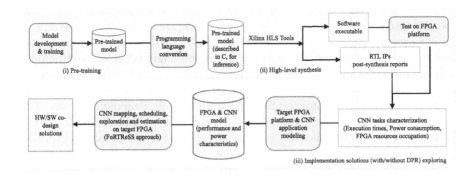

Fig. 1. Overview of our approach workflow. Firstly, we perform training using Keras to process the weights and bias values of the CNN model. Secondly, with the help of HLS tools, we synthesize our CNN model described in C to get the RTL IPs, which subsequently enable us to measure the inference time, power, and resource quantity. These characteristics allow the modeling of target FPGA and CNN layers, which are necessary to explore the implementation solutions using FoRTReSS approach.

To identify relevant implementation solutions in this very large design space, it is crucial to consider execution time and resource consumption of each task, but also power, energy, and reconfiguration times. Hardware implementations are defined by considering actual task resources and performance, which can be based on HLS and lower-level estimates from FPGA synthesis tools. The target architecture is described by detailed FPGA resource organization and a set of processor cores. Using the characterization of the application tasks and the target platform architecture, FoRTReSS can automatically explore all possible RRs and finally produce a set of mappings fully scheduled on different cores and RRs. FoRTReSS provides energy consumption estimates for the full application and for software and hardware implementations of each task as well, allowing decisions on optimal energy efficiency at different levels.

In the following, we introduce the approach for modeling CNN applications and target platforms, as well as the exploration of design space by FoRTReSS. Specifically, in Sect. 3.2 and Sect. 3.3, we provide an example showing the characterization and exploration results of a GTSRB (German Traffic Sign Recognition Benchmark, 32 × 32 color traffic sign images) CNN topology.

3.1   High-Level Synthesis

The definition and simulation of the original CNN topology, also producing learning weights and bias values, are processed using Keras. Pure synthesizable C code has been developed to help fast prototyping of ConvNets and comprehensive hardware/software deployment on FPGA following an HLS methodology of Xilinx. A 16-bit quantization scheme (Q8.8) is first applied currently to the reference 32-bit floating-point weight and bias values to produce a fixed-point model. The subsequent layer-based implementation supports full hardware implementation of 2D convolutional, max-pooling, and fully connected layers. Other processing functions such as file reading, pre-processing and post-processing of images are kept in software. This approach allows the creation of a wide range of popular network topologies, with quasi-automated RTL design ensuring complete practical implementation and execution on real FPGA platforms. Intra-layer parallelism is explored in terms of loop unrolling, pipelining, and array partitioning.

For each layer of the network, we explore loop-level parallelism which remains a manual process in the RTL design flow. Within the entire network architecture, convolutional layers are the most computationally demanding layers. The 2D convolution is based on six nested loops processing output channels, output feature maps (rows and columns), input channels, and K*K convolution kernels. Vivado HLS offers different pragmas to optimize the scheduling and resource allocation within these nested loops. A pipeline directive at the input channel level gives the best tradeoff between performance and on-chip resources. The inner nested loops are automatically unrolled by Vivado HLS, provided concurrent read/write accesses to the different arrays are possible. The input arrays are therefore partitioned adequately to let them to be stored in different BRAMs.

3.2   System Modeling

All the measurements and estimations described in the following constitute the necessary elements of an abstract model of the complete system, which can be further used by FoRTReSS DSE to process fast and reliable CNN mapping analysis and scheduling simulations.

Layer-Wise Processing Functions: For the CNN model in FoRTReSS, each layer is treated as an individual task, which may have a hardware or a software implementation. For hardware implementations, area and power are estimated for each task using Xilinx post-synthesis estimators. The power estimated for hardware implementations includes two components: idle and running power. Idle power represents the power needed by a task when configured in a RR but not actively executing. Running power refers to the additional power consumed during task execution within the RR. Execution times are derived from Vivado HLS reports. For software implementations, power characterization of a task is based on the power model of the Processing System (PS) of the platform. Execution times are derived from real measurements on one CPU core.

Reconfigurable Platform: This work targets Zynq series, specifically with a ZedBoard (dual-core Arm Cortex-A9, XC7Z020-CLG484-1) from the Zynq-7000 family and a ZCU102 (quad-core Arm Cortex-A53, XCZU9EG-FFVB1156-1) from the Zynq Ultrascale+ family, both of which support Dynamic Partial Reconfiguration (DPR). For the target reconfigurable platforms considered, modeling is usually divided into two distinct sections: the PS and the Programmable Logic (PL). The power model for PS is composed of three parts: static, idle, and active. The construction of this power model refers to the processor power model presented in [8], which can be generally expressed by the following formula:

$$P_{cpu} = P_{cpu}^{static} + \alpha * P_{cpu}^{idle} + \beta * P_{cpu}^{run} \tag{1}$$

where α and β are the coefficients expressing the idle and running contribution respectively for a given type of CPU.

Concerning the PL, we model the amount and arrangements of the different resources (logic cells, RAM blocks, DSP blocks) of the entire FPGA using XML files. For the set of RRs generated by FoRTReSS, the power model is similarly divided into three components: empty, idle, and running [9]. The empty (static) power corresponds to the power consumption of a region when it is unoccupied, which is estimated proportionally to the full static power consumption of the FPGA device, depending on the FPGA resources within the region. Idle and running power are associated with the task when it is configured and executed, respectively, on the corresponding region. Each component of the power model mentioned above is derived from Xilinx Power Estimator after logic synthesis.

Additionally, when implementing hardware with dynamic reconfiguration, it is fundamental to address reconfiguration efficiency. Reconfiguration times are excessive using original Xilinx controllers (ICAP, PCAP) which affects both performances and energy very significantly. Our approach to this well-known DPR problem is based on using the currently known best-performance reconfiguration controller in the literature, UPaRC (Ultra-fast Power-aware Reconfiguration Controller [10]), which can reach a reconfiguration throughput of 1.433GB/s. The reconfiguration controller is mainly characterized by two parameters, P_{reconf} and T_{reconf}, which characterize respectively the associated reconfiguration power and performance. P_{reconf} is extrapolated from [10] to consider the maximum operating frequency of 362.5 MHz (thus estimated at 460 mW). It is worth noting that due to its relatively small size (around 1000 slices), the associated idle power is set to zero. Reconfiguration time is estimated as follows:

$$T_{reconf} = \frac{Bitfile\ size}{Reconf\ controller\ throughput} \tag{2}$$

where the size of the bitfile can be calculated by FoRTReSS, by evaluating the quantity of resources of the RR. Similarly to the hardware static power, the reconfiguration time of a region depends on the size of that region, implying that smaller reconfiguration regions should be selected.

Application to GTSRB/ZedBoard: The following details an example of CNN characterization. The application is a seven-layer convolutional neural network targeting GTSRB on ZedBoard. Key features of the implementation model include task id i, execution unit j (core or RR), execution time $T_{i,j}$, power $P_{i,j}$, and the number of FPGA resources N^{slice}, N^{dsp}, and N^{bram} for hardware implementations. Each task encompasses one or more implementations, with at least one software implementation.

As illustrated in Table 1, the hardware acceleration effect is significant for convolutional layers, especially for conv1 and conv3. In cases where energy consumption is also advantageous and depending on other scheduling constraints, these hardware implementations will be considered in the first place for execution. Conversely, for layers with approximately equal execution time on software and hardware (pool1, pool2) or faster execution on software (fc1, fc2), also considering the reconfiguration overhead, the allocation would be made preferably on a software core.

Table 1. Software/hardware characterization of GTSRB application tasks on Zedboard

Task (i)	Execution unit (j)	$T_{i,j}$ (ms)	$P_{i,j}^{static}$/ $P_{i,j}^{idle}$/$P_{i,j}^{run}$ (mW)	N^{slice}/N^{bram}/N^{dsp}
img_load	Core	0.730	45 / 8 / 257.7	–
conv1(5 × 5, 6)	Core	2.247	45 / 8 / 257.7 – / 42 / 67	– 1462 / 0 / 25
	RR	0.898		
maxpooling1	Core	0.023	45 / 8 / 257.7 – / 41 / 2	– 90 / 0 / 0
	RR	0.024		
conv2(5 × 5, 16)	Core	1.334	45 / 8 / 257.7 – / 42 / 62	– 450 / 2 / 25
	RR	1.285		
maxpooling2	Core	0.009	45 / 8 / 257.7 – / 41 / 3	– 78 / 0 / 0
	RR	0.008		
conv3(5 × 5, 120)	Core	0.331	45 / 8 / 257.7 – / 41 / 47	– 256 / 0 / 15
	RR	0.066		
fc1(84)	Core	0.066	45 / 8 / 257.7 – / 41 / 2	– 30 / 0 / 1
	RR	0.105		
fc2(43)	Core	0.025	45 / 8 / 257.7 – / 41 / 2	– 21 / 0 / 1
	RR	0.041		
softmax	Core	0.010	45 / 8 / 258	–

3.3 Design Space Exploration

System Scheduling: FoRTReSS generates and simulates full system mappings and associated schedulings for various possible deployments of the application /platform model. An illustrative example of this is provided in Fig. 2, which presents one implementation solution for the GTSRB topology using two RRs.

This scheduling trace illustrates the detailed and actual allocation, execution, and reconfiguration of each individual task. It additionally shows power and energy consumption at each scheduling event, for possibly several hyperperiods. Table 2 details values of energy, performance, and power per each CNN layer, from GTSRB PR scheduling simulation with two RRs on ZedBoard.

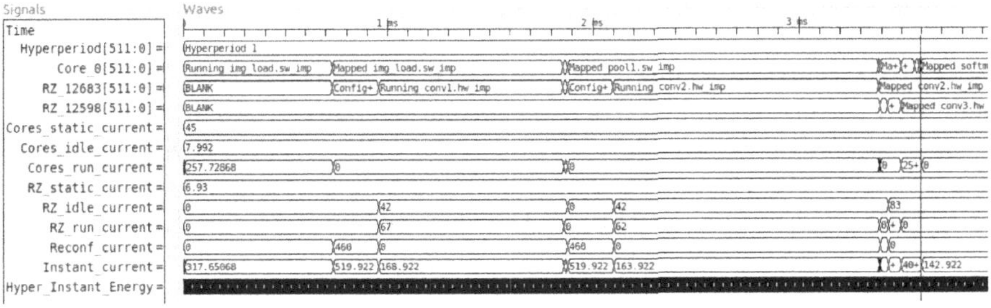

Fig. 2. Scheduling simulation results for GTSRB PR execution with two RRs

RR Partitioning: Fig. 3 illustrates the placement of two RRs as part of a PR implementation solution for the GTSRB topology on Zedboard. FoRTReSS defines several candidates RRs based on the resources available on the device and the characterization of tasks. To make the best choices, these RRs are further evaluated using a cost function. This takes into account how complex their shape is, how many tasks they can handle, and how much of their resources are left unused, known as internal fragmentation [11]. This careful evaluation helps in defining and picking the most relevant regions for the tasks and identifying the best FPGA partitioning for the global application.

4 Validation Study

In the following, we address the design space exploration of three typical CNN benchmarks by comparing their software and hardware implementations (static and PR): MNIST (Modified National Institute of Standards and Technology database, 28×28 grayscale handwritten digits), GTSRB and CIFAR-10 (Canadian Institute for Advanced Research, 32×32 color object and animal images). We consider three standard topologies associated with these benchmarks, that are further trained and validated using Keras to serve as a reference for accuracy and provide the weights and biases used for CNN inference in the corresponding HLS C code. For the topologies associated with MNIST and GTSRB, we analyze different implementations (software, static hardware, PR hardware) on ZedBoard using FoRTReSS. For CIFAR-10 we consider a ZCU102 platform since ZedBoard is too small. For PR hardware, we only report the best energy-efficient solution with the optimal number of RRs.

On top of these three benchmarks, we also address a short comparison with existing works related to LeNet-5 on XC7Z020, in a way to better assess the relevance of numbers and conclusions reported.

4.1   CNN Benchmarks Exploration Results

MNIST: The first benchmark is a five-layer CNN applied to the MNIST dataset (improved LeNet-5), composed of two convolutional layers, two max-pooling layers, and two fully connected layers (28 × 28-20C5-P2-40C5-P2-400-10). The corresponding classification accuracy is 99.05%. The best energy-efficient solution for PR HW identified by FoRTReSS DSE is based on using two RRs.

For the MNIST dataset, both hardware implementations perform better in energy consumption than software (Table 3). PR HW reports 2.59 times faster and 1.75 times better energy efficiency compared to software. However, due to reconfiguration times, PR is 2.03 times slower than static hardware, particularly for the conv2 layer, as the associated RR is quite large, covering nearly half the size of the FPGA device. The corresponding reconfiguration time ends up representing around 25% of the total CNN inference time, also negatively impacting the energy consumption of the PR solution which is 1.98 times less energy-efficient than static hardware.

Table 2. Power and performance breakdown for GTSRB PR execution on Zedboard

Task (i)	Energy (mJ)	Execution time (us)	P_core (mW)[1]	P_RZ (mW)[1]	P_Reconf (mW)	P_total (mW)
img_load	0.230	730	45 + 8 + 257.7	6.93 + 0 + 0	0	317.7
config_conv1	0.116	222 (RZ 1)	45 + 8 + 0	6.93 + 0 + 0	460	519.9
running_conv1	0.152	898	45 + 8 + 0	6.93 + 42 + 67	0	168.9
running_pool1	0.007	23.1	45 + 8 + 257.7	6.93 + 42 + 0	0	359.7
config_conv2	0.117	222 (RZ 1)	45 + 8 + 0	6.93 + 0 + 0	460	519.9
running_conv2	0.211	1285	45 + 8 + 0	6.93 + 42 + 62	0	163.9
running_pool2	0.002	8.98	45 + 8 + 257.7	6.93 + 42 + 0	0	359.7
config_conv3	0.022	38.3 (RZ 2)	45 + 8 + 0	6.93 + 42 + 0	460	561.9
running_conv3	0.013	66	45 + 8 + 0	6.93 + 83 + 47	0	189.9
running_fc1	0.026	65.9	45 + 8 + 257.7	6.93 + 83 + 0	0	400.7
running_fc2	0.010	24.9	45 + 8 + 257.7	6.93 + 83 + 0	0	400.7
softmax	0.004	10	45 + 8 + 257.7	6.93 + 83 + 0	0	400.7
Total	**0.910**	**3594**	-	-	-	-

[1] P = P_static + P_idle + P_run

It is worth noting that in terms of average power, and despite previous net energy benefits, both static and PR HW consume more than software (468.5 mW and 458.8 mW, vs. 310.7 mW). This is due to the relatively high static power consumption of FPGA devices compared to hard IP cores on Zynq. In terms of resource occupation, Table 4 reports a slight increase of Slices and DSPs for PR HW, and around 2x more BRAM compared to static HW. This can be attributed to the fact that the total resources covered by the rectangular RR generated for conv2 marginally exceed those of the static implementation.

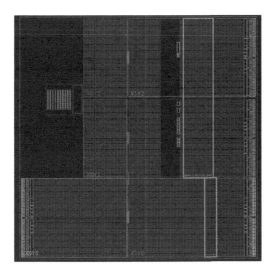

Fig. 3. Layout of RRs placement for the solution HW PR with two RRs for GTSRB topology on Zedboard

Table 3. Efficiency comparison of software and hardware (static and PR) implementations for MNIST, GTSRB (Zedboard) and CIFAR-10 (ZCU102)

Implementation	Execution time (ms)	Average Power (mW)	Energy consumption (mJ)
MNIST SW	11.565	310.7	3.593
MNIST Static HW	**2.205**	468.5	**1.033**
MNIST PR HW	4.466	458.8	2.049
GTSRB SW	4.775	310.4	1.482
GTSRB Static HW	**3.112**	357.3	1.112
GTSRB PR HW	3.594	256.5	**0.910**
CIFAR-10 SW	198.697	220.1	43.734
CIFAR-10 Static HW	**14.078**	1341.2	18.882
CIFAR-10 PR HW	16.535	682.0	**11.277**

GTSRB: The topology used with the GTSRB dataset consists of three convolutional layers, two max-pooling layers, and two fully connected layers (32×32-6C5-P2-16C5-P2-120C5-84-43). Classification accuracy is 85.20% for this benchmark. The best energy-efficient solution for PR HW identified by FoRTReSS DSE is based on using two RRs.

Like previously on the MNIST dataset, hardware solutions have better performance and energy efficiency compared to software execution (Table 3). PR HW brings 1.33 times (24.7%) better inference time and 1.63 times (38.6%) higher energy efficiency compared to software. In terms of execution time, PR HW remains slightly slower (+15.5%) than static HW due to the additional reconfiguration times.

Table 4. Resource occupation of hardware implementations (static and PR) for MNIST, GTSRB (Zedboard) and CIFAR-10 (ZCU102)

Implementation	Functions on HW	N^{slice}	N^{bram_18k}	N^{dsp}
MNIST Static HW	conv1, conv2, fc1	**6764** / 13300	99 / 280	136 / 220
MNIST PR HW (2 RRs)		6900 / 13300	220 / 280	140 / 220
GTSRB Static HW	conv1, conv2, conv3	2168 / 13300	2 / 280	65 / 220
GTSRB PR HW (2 RRs)		**1900** / 13300	20 / 280	60 / 220
CIFAR-10 Static HW	conv2, conv3, conv4, conv5, conv6	28408 / 34260	6 / 1824	777 /2520
CIFAR-10 PR HW (2 RRs)		**18360** /34260	48 / 1824	480 / 2520

However, in contrast to the MNIST benchmark, PR HW is 1.22 times (18.2%) more energy-efficient than static implementation. With more layers, there is also more potential for hardware acceleration and better efficiency. In addition, for this benchmark PR HW saves more resources compared to static HW (Table 4), resulting in a more significant improvement of 28% in average power consumption (256.5 mW vs. 357.3 mW). Comparing with software (310.4 mW) further highlights the benefits of PR on this example, but also the relatively high cost of static power inherent to static HW implementations.

CIFAR-10: The last benchmark, which is also the largest topology, consists of six convolutional layers, three max-pooling layers, and one fully connected layer (32 × 32-32C3-32C3-P2-64C3-64C3-P2-128C3-128C3-10). It is applied to the CIFAR-10 dataset with a classification accuracy of 79.2%. The best energy-efficient solution for PR HW identified by FoRTReSS DSE utilizes two RRs.

For this dataset, as demonstrated in Table 3, PR HW continues to report faster performance and lower energy consumption than software. It is notably 12.02 times (91.7%) faster and 3.88 times (74.2%) more energy-efficient, indicating a more pronounced improvement compared to the two previous topologies. Similar to the GTSRB benchmark, the PR implementation is slightly slower (+17.5%) than static HW due to reconfiguration overheads.

It should be noted here that the CIFAR benchmark, the largest CNN example in this application study, is where PR has the best results against static implementation with 1.67 times (40.3%) more energy efficiency. In this benchmark, PR HW provides the most important FPGA resource savings among the three benchmarks examined as well (Table 4). Similar to MNIST and GTSRB, PR HW occupies more BRAM resources than static HW. This is primarily due to the fact that the tasks within this topology utilize fewer BRAMs. However, FoRTReSS generates rectangular RRs that occupy a full column of BRAMs. As a consequence, additional BRAMs are present and contribute to the total increased count. Conversely, it achieves approximately 40% savings for Slice and DSP resources. This results in better average power consumption for PR against static HW with 49% improvement.

Unlike GTSRB, here PR HW is less power efficient than software (682.0 mW vs. 220.1 mW). This is due to the size of the network, featuring more layers, and leading to increased resource usage and higher static power in comparison to GTSRB. Static HW is not very efficient for the same reason, consuming up to six times more average power than software for this relatively large topology.

In light of these results, it appears that PR can notably improve the processing efficiency of static hardware acceleration for CNNs, provided certain conditions are met. This is clearly the case for the two largest topologies investigated, GTSRB and CIFAR-10, with respectively 1.22 (18.2%) and 1.67 times (40.3%) more energy efficiency than static hardware. However, for MNIST which is the smallest topology, fewer layers means less hardware acceleration and RR reuse potential, making PR improvements less impactful.

4.2   Comparison with Other Works

In this section, we attempt to compare our results to other works. It should be noted here that it is difficult to identify some works for this, with so many design options: CNN benchmark and topology, FPGA device, and implementation parallelism. Most works report on actual performances, among which few allow for power and energy comparison, also considering both static and dynamic solutions. In the following, we have been able to set up an interesting comparison of the MNIST benchmark with fpgaConvNet [6], in terms of performance. Indeed the paper reports a measured performance of 0.48 GOp/s for a LeNet-5 CNN (32×32-6C5-P2-16C5-P2-120-84-10, 98.57%) on an XC7Z020 device at 100 MHz. [13] (2019) is another work addressing the implementation of LeNet-5, but there is no PR involved. However, they report comparable performance with 0.343 GFLOPS on the ZYBO Z7 board (XC7Z020) at 100 MHz.

Our results on the MNIST (28×28-20C5-P2-40C5-P2-400-10, 99.05%) on an XC7Z020 device at 100 MHz, despite higher topology complexity, reflect a similar level of performance with 0.83 GOp/s for the static implementation and 0.41 GOp/s for the dynamic implementation. In other words, we are at the same level of performance with [6] but for a twice bigger topology with approximately 50% fewer resources for the dynamic solution (cf. Table 4). This would allow to target smaller devices for implementation, thus highly reducing power consumption when considering that MNIST is the least energy efficient of the three benchmarks investigated.

5   Conclusion and Perspectives

In this work, we proposed a novel and elegant approach to improve the efficiency of CNN inference on FPGAs exploiting reductive PR. This approach partitions the original CNN topology based on various layers and reconfigures the PL with a smaller number of reconfigurable resources at runtime. This methodology greatly helps to explore automatically different FPGA partitioning and arrangements of

RRs and to identify more energy-efficient hardware/software mapping solutions given the extended size of the PR codesign space.

We assessed this approach on three different network examples, showing potential improvements reaching a factor of 3.88 (gain vs. software solution) and 1.67 (gain vs. static solution) in energy savings. The improvement levels greatly depend on the inner potential of layers for hardware acceleration and grow steadily with the depth of the network as more hardware reuse is possible.

From these results, future works will investigate an application study on deeper, widely-used CNN networks, such as ResNet-50. Ongoing works also address the automated implementation of the identified solutions on real platforms following the methodology of [12] on Xilinx Zedboard and ZCU102 platforms.

References

1. Farhadi, M., Ghasemi, M., Yang, Y.: A novel design of adaptive and hierarchical convolutional neural networks using partial reconfiguration on FPGA. In: IEEE High Performance Extreme Computing Conference (HPEC), Waltham, MA, USA, pp. 1-7 (2019). https://doi.org/10.1109/HPEC.2019.8916237.
2. Youssef, E., Elsimary, H.A., El-Moursy, M.A., Mostafa, H., Khattab, A.: Energy-efficient precision-scaled cnn implementation with dynamic partial reconfiguration. IEEE Access 10, 95571–95584 (2022). https://doi.org/10.1109/ACCESS.2022.3204704
3. Youssef, E., Elsemary, H.A., El-Moursy, M.A., Khattab, A., Mostafa, H.: Energy Adaptive convolution neural network using dynamic partial reconfiguration. in: IEEE 63rd International Midwest Symposium on Circuits and Systems (MWSCAS), vol. 2020, pp. 325–328. Springfield, MA, USA (2020). https://doi.org/10.1109/MWSCAS48704.2020.9184640
4. Irmak, H., Ziener, D., Alachiotis, N.: Increasing Flexibility of FPGA-based CNN Accelerators with Dynamic Partial Reconfiguration. In: 2021 31st International Conference on Field-Programmable Logic and Applications (FPL), Dresden, Germany, 2021, pp. 306-311. https://doi.org/10.1109/FPL53798.2021.00061.
5. Alberto de Albuquerque Silva, C., Andrey Ramalho Diniz, A., Duarte Dória Neto, A., Alberto Nicolau de Oliveira, J.: Use of partial reconfiguration for the implementation and embedding of the artificial neural network (ANN) in FPGA. In: 4th International Conference on Pervasive and Embedded Computing and Communication System (2014). https://doi.org/10.5220/0004716301420150.
6. Venieris, S.I., Bouganis, C.S.: fpgaConvNet: a framework for mapping convolutional neural networks on FPGAs. In: IEEE International Symposium on Field-Programmable Custom Computing Machines (FCCM), Washington, DC, USA, vol. 2016, pp. 40–47 (2016). https://doi.org/10.1109/FCCM.2016.22
7. Kastner, F., Janben, B., Kautz, F., Hubner, M., Corradi, G.: Hardware, software codesign for convolutional neural networks exploiting dynamic partial reconfiguration on PYNQ. In: IEEE International Parallel and Distributed Processing Symposium Workshops (IPDPSW), Vancouver, BC, Canada, vol. 2018, pp. 154–161 (2018). https://doi.org/10.1109/IPDPSW.2018.00031
8. Bonamy, R., et al.: Energy efficient mapping on manycore with dynamic and partial reconfiguration: application to a smart camera. Inter. J. Circuit Theory Appli. (2018). https://doi.org/10.1002/cta.2508

9. Bonamy, R., Bilavarn, S., Chillet, D., Sentieys, O.: Power modeling and exploration of dynamic and partially reconfigurable systems. J. Low Power Electr. (2016). https://doi.org/10.1166/jolpe.2016.1448

10. Bonamy, R., Pham, H.-M., Pillement, S., Chillet, D.: UPaRC-Ultra-fast power-aware reconfiguration controller. In: Design, Automation & Test in Europe Conference & Exhibition (DATE), Dresden, Germany, vol. 2012, pp. 1373–1378 (2012). https://doi.org/10.1109/DATE.2012.6176705

11. Duhem, F., Muller, F., Bonamy, R., Bilavarn, S.: Fortress: a flow for design space exploration of partially reconfigurable systems. Design Autom. Embedded Syst., 301-326 (2015). https://doi.org/10.1007/s10617-015-9160-2.

12. Sadek, A., et al.: Supporting utilities for heterogeneous embedded image processing platforms (STHEM): an overview. In: International Symposium on Applied Reconfigurable Computing (ARC) (2018). https://doi.org/10.1007/978-3-319-78890-6_59

13. Rongshi, D., Yongming, T.: Accelerator implementation of Lenet-5 convolution neural network based on FPGA with HLS. In: 2019 3rd International Conference on Circuits, System and Simulation (ICCSS), Nanjing, China, pp. 64-67 (2019). https://doi.org/10.1109/CIRSYSSIM.2019.8935599.

Optimising Graph Representation for Hardware Implementation of Graph Convolutional Networks for Event-Based Vision

Kamil Jeziorek[1] , Piotr Wzorek[1] , Krzysztof Blachut[1] , Andrea Pinna[2] ,
and Tomasz Kryjak[1,2(✉)]

[1] Embedded Vision Systems Group, Department of Automatic Control and Robotics,
AGH University of Krakow, Krakow, Poland
{kjeziorek,pwzorek,kblachut,tomasz.kryjak}@agh.edu.pl
[2] Sorbonne Université, CNRS, LIP6, 75005 Paris, France
andrea.pinna@lip6.fr

Abstract. Event-based vision is an emerging research field involving processing data generated by Dynamic Vision Sensors (neuromorphic cameras). One of the latest proposals in this area are Graph Convolutional Networks (GCNs), which allow to process events in its original sparse form while maintaining high detection and classification performance. In this paper, we present the hardware implementation of a graph generation process from an event camera data stream, taking into account both the advantages and limitations of FPGAs. We propose various ways to simplify the graph representation and use scaling and quantisation of values. We consider both undirected and directed graphs that enable the use of PointNet convolution. The results obtained show that by appropriately modifying the graph representation, it is possible to create a hardware module for graph generation. Moreover, the proposed modifications have no significant impact on object detection performance, only 0.08% mAP less for the base model and the N-Caltech data set. Finally, we describe the proposed hardware architecture of the graph generation module.

Keywords: graph representation · GCN · event cameras · object detection · FPGA

The work presented in this paper was supported by: the AGH University of Krakow project no. 16.16.120.773, the program "Excellence initiative – research university"' for the AGH University of Krakow, Polish high-performance computing infrastructure PLGrid (HPC Centers: ACK Cyfronet AGH) – grant no. $PLG/2023/016130$ and partly by Sorbonne University. Tomasz Kryjak would like to sincerely thank Sorbonne Université for inviting him and funding his stay as a visiting professor in November 2022.

T. Dias and P. Busia (Eds.): DASIP 2024, LNCS 14622, pp. 110–122, 2024.
https://doi.org/10.1007/978-3-031-62874-0_9

1   Introduction

Event cameras, also known as neuromorphic cameras, are an emerging type of a video sensor. Unlike standard frame cameras, which read out the brightness values of all pixels simultaneously in a fixed time interval, event cameras record the brightness changes of individual pixels independently. This allows correct operation in challenging lightning conditions, such as poor illumination and high dynamic range, and reduces average power consumption. In addition, information on pixel brightness changes is obtained with a much higher temporal resolution – in the order of microseconds vs 16 millisecond for a typical 60 frame per second frame camera.

One promising application of event cameras is the detection and tracking of moving objects, such as vehicles and pedestrians. The use of neuromorphic sensors can significantly improve safety at key moments. However, due to their asynchronous nature of operation, event cameras generate an irregular data stream in the form of a sparse spatio-temporal event cloud, which is difficult to process. The most commonly proposed solution is to create pseudo-frames that mimic traditional video frames or to use event-based video frame reconstruction. Both approaches produce good results, but lead to the loss of the advantages of high temporal resolution.

Therefore, recent research has focused on processing event data in its original sparse form. An example is the use of spiking neural networks (SNNs) [3]. Inspired by biology, it is a promising approach but suffers from a lack of standardised learning algorithms, making it difficult to apply or achieve high performance in more complex methods. Another recently proposed solution is the use of Graph Convolutional Networks (GCNs), which allow event data to be processed in the form of a spatio-temporal point cloud. In addition, recent work in this area suggests that such representations can be updated asynchronously [5,7,10]. However, current computations are mainly performed using high-energy GPUs, and thus similar architectures cannot be used in low-power embedded vision systems. Furthermore, to the best of our knowledge, current research in data processing with GCNs relies primarily on pregenerated graphs or specifies their construction for a pre-established structure.

In this paper, we present the hardware implementation of a graph generation module for event data on SoC FPGAs (System-on-Chip Field Programmable Gate Arrays). Our primary contributions include the development of a graph representation that takes into account the architectural strengths and limitations of SoC FPGAs, leading to a reduction in the *number of graph edges* while still achieving detection and classification results comparable to recent methods. Additionally, we introduce the concept of an FPGA hardware module for *graph representation generation*, which produces a representation ready for processing by the selected PointNet convolution layer [9].

The remainder of the paper is organised as follows: Sect. 2 presents the issue of object detection using graph convolutional networks. Section 3 describes the methods we have considered to optimise the graph representation of event data. In Sect. 4 we present the concept of hardware-based graph generation on FPGAs.

The article concludes with a summary 5 of the results obtained and a presentation of plans for further work.

2   Object Detection Using GCN

2.1   Graph Construction

For an event camera, the event generation process is controlled by a threshold mechanism. An event is generated when the logarithmic change in light intensity (ΔL) at a particular pixel (x_i, y_i) and time t_i exceeds a fixed threshold value (C) for the same pixel at time t_i^*: $\Delta L(x_i, y_i, t_i) = L(x_i, y_i, t_i) - L(x_i, y_i, t_i^*) \geq p_i C$.

Each generated event includes: the coordinates of the pixel, the time received from the internal camera clock, and the polarisation p_i in the range $\{-1, 1\}$, specifying the direction of the change in light intensity. These four values form a tuple $e_i = \{x_i, y_i, t_i, p_i\}$, and the events together form a data sequence $E = \{e_i\}_{i=0}^{N-1}$ (N – number of events).

The graph structure $\mathcal{G}$ consists of a set of *vertices* $\mathcal{V}$ and a set of *edges* $\mathcal{E}$ that reflect the relationships between vertices. A commonly accepted method for event-based graph generation is to treat each event as a vertex v with spatio-temporal position $pos = (x, y, t)$ and attribute $a = (p)$. Edges are created based on a chosen distance metric between vertex positions, most commonly using an Euclidean distance bounded by a radius of length R. A full description of the graph representations used is given in Sect. 3.

2.2   Graph Convolutional Networks

Graph Convolution. Currently, there are many different convolution operations on a graph, differing in the information used, the functions and the type of data processed. Recent work in this area apply operations such as SplineConv or GCNConv. However, the study [6] showed that the simpler and memory-efficient PointNetConv operation [9] can provide comparable results in event data detection and classification problems. We therefore decided to use PointNetConv and model presented in [6] as a reference in our work, as it is a better candidate for embedded implementation.

The graph convolution operation for a vertex v_i in a graph $\mathcal{G}$ consists of three steps: a *message function* that utilises the attributes of a vertex a_i, as well as those of its neighbours a_j and the difference in vertex positions $pos_j - pos_i$, which are transformed by the linear function h_Θ. The *aggregation function* then selects the maximum value among all messages for a given vertex v_i. Finally, the *update function* applies an additional linear function h_Θ to the output of the aggregation function to adjust the output size.

These steps are performed for each vertex in the graph $\mathcal{G}$, and at each successive convolution layer, the attribute dimension a_i of the vertices is modified, with the spatio-temporal position unchanged.

Graph Pooling. The pooling process, relevant in the context of GCNs, involves the selection of representative vertices or the aggregation of information within a group or cluster. Selection criteria may include the relevance of the vertex, the maximum or average value for all vertices in the group. The result of this process is a new vertex that represents the entire group.

In the context of GCNs, the pooling process enables the creation of denser representations, which translates into a reduction of computational complexity in subsequent layers. This encourages more efficient processing of large-scale data, while preserving essential features and smooth information flow. It is worth noting that this operation also enables the use of fully connected layers data classification.

2.3   Related Work

Due to the lack of work that focuses on combining the potential of GCNs, event-based data and hardware implementation on FPGAs, we start the review with methods that use GCNs to process event-based data. We then move on to analyse work that focuses on using FPGAs to accelerate GCNs.

Object Detection Using GCN for Event Data. Initially, GCNs were used to process event data because they allowed asynchronous processing at the individual event level. One of the first work [11] proposed an EventNet architecture that processed events in a recursive manner using temporal encoding. Subsequent work further developed this area, enabling object recognition based on event data. In the work [2], residual graph blocks were used, while the work [7] introduced asynchronous graph construction and update using slide convolution. Recent research in this area [5,10] has shown that object detection is also possible using GCNs. Asynchronicity was achieved by updating only neighbouring vertices in each successive layer, reducing the computational cost and latency of processing each event. A different approach, compared to previous works, was presented in [6]. The focus was on developing memory-efficient GCNs for event data processing, utilising basic convolutional layers tailored for point-cloud data processing.

Despite these advances, all of the work mentioned is mainly based on high-energy platforms such as GPUs. This makes it impossible to develop energy-efficient detection systems using event cameras.

Graph Acceleration on FPGA. Implementations of GCNs on FPGAs primarily concentrate on hardware acceleration of the networks themselves. The paper [12] introduces a lightweight GCN accelerator on an FPGA, while [1] presents an accelerator tailored to various types of convolutional layers. Additionally, the work [4] explores network acceleration on multi-FPGA platforms.

It is worth noting, however, that much of the existing work is centred on processing established graphs or generating graphs based on predefined structures, such as the outputs of accelerator sensors [8]. In contrast, when dealing with event data characterised by a random spatio-temporal nature, the required approach differs. It this work, we try to address the latter issue.

3   The Proposed Graph Representation Optimisation

3.1   Graph Representations

A variety of graph structure representation methods can be found in the literature, such as Compressed Sparse Row (CSR) and Compressed Sparse Column (CSC). Nevertheless, the most popular format is the Coordinate (COO) format.

In COO format, the edges are represented by vectors containing source and destination vertex indices. In GCNs implementations using off-the-shelf libraries, such as PyTorch Geometric, as demonstrated in works [5,6,10], this representation format is often used because of its simplicity.

Since the number of potentially generated edges linked to an event can vary significantly, initialising sufficient memory to process and store the data in an FPGA is impossible. To overcome this hardware limit, we propose our own modified representation.

3.2   Optimisation

First, it is crucial to determine whether the graph will consist of directed or undirected edges. In our approach, we establish that with the arrival of each new vertex, previous events are directed towards this new event, indicating that the graph is directed based on time. This procedure enables us to apply updates solely to the newly examined vertex during the convolution operation.

As we intend to use the PointNet convolution in our work, it becomes necessary to determine the attributes of a vertex, its neighbours, and the positional difference between vertices connected by edges. To achieve this, for each vertex, we record its own position $pos = (x, y, t)$, the attribute $a = (p)$ (polarity), and generate edges that specify the connected vertices. For this purpose, we introduce three optimisation methods.

Normalisation. The locations values of the incoming events depend on the matrix resolution, while the time is represented in microseconds. In existing works [5,6,10], the normalisation only applies to time, where it is multiplied by some factor to make it close to the dimension of the x and y values. Instead, we propose normalising the spatial and temporal dimensions to a uniform range ($SIZE$). In addition, we quantise the time after normalisation to an integer value. This results in a spatio-temporal representation where event values are in the range $(0, 0, 0) - (SIZE, SIZE, SIZE)$.

Table 1. Average number of nodes and edges for data normalised to a given graph size. Methods selected for hardware implementation are in **bold**.

Graph size	Event Preprocessing			Edge Generation		Average Nodes	Average Edges
	Without	Random	Unique	Radius	NM		
128	x			x		56,940	836,304
	x				x		309,711
		x		x		25,000	173,868
		x			x		70,896
			x	x		**50,498**	623,078
			x		x		**262,935**
256	x			x		56,940	138,757
	x				x		67,043
		x		x		25,000	28,892
		x			x		14,051
			x	x		**55,925**	131,648
			x		x		**65,214**

Neighbour Matrix. In existing solutions, edge generation requires searching for all vertices or a relevant subgroup to identify potential neighbours. Implementing this effectively in an FPGA is difficult due to the requirement of sorting a large number of vertices. To address this problem, we propose a two-dimensional neighbour matrix (NM) for edge generation, which has dimensions of $SIZE \times SIZE$ after normalisation. When a new event with position (x, y, t) occurs, we check its local neighbourhood defined by the search radius R in NM. This approach allows us to search for neighbours only in the group of nearest values, while reducing the number of potentially generated edges. We also avoid the situation where one vertex is connected by edges to several vertices with the same spatial position (x, y) but different times t. Only the most recent vertex from a given pixel is considered.

Unique Events. If, for a given pixel (x, y) in the neighbour matrix, a vertex with the time t already exists, we treat the appearance of an event with the same time as a duplicate and do not include it in the edge generation process.

By using these methods, the maximum number of potential vertices in the entire graph, which depends on the $SIZE$ parameter, can be determined. The number of generated edges depends on the search context, defined by the radius R. Due to the use of time-directed edges, vertices with generated edges can be stored in external memory, eliminating the need for their later updates.

3.3   Ablation Studies

To examine the impact of the optimisations considered, we performed ablation studies on the N-Caltech101 dataset[1]. From each sample, we extracted a 50 ms time window with the highest event density. The ablation studies were divided into two main components: the effect of the methods on the size of the generated graph and the training results of the model considered. For each of these elements, we distinguished three subgroups of influencing methods: value normalisation, event preprocessing, and edge generation method.

The methods we presented were compared with the solutions proposed in [10], in which the number of events is sampled at *random* to a fixed value (we selected the number to be 25,000) and using the *radius* method, in which up to 32 neighbours are searched within a distance R (we selected R to be equal to 3), considering all vertices in the graph. We also compared the methods *without* taking into account the preprocessing of events.

Impact of the Methods on the Size of the Generated Graph. We noticed the biggest difference in the number of generated edges between *radius* and the NM (neighbour matrix) methods (Table 1). This is mainly due to the fact that the *radius* function generates undirected edges. Additionally, when using NM, we avoid edge multiplications within the same pixel that occur in a short time interval. Normalisation to 128 means the reduction of around 12% of the vertices, while for normalisation to 256 the reduction is only 2%. A separate analysis of the average pixel activity showed that the use of unique values in particular reduced the impact of the occurrence of overly bright pixels. Although the use of *random* preprocessing significantly reduces the number of vertices and edges, dropping events to a fixed value without prior knowledge of how many there will be is impossible. This shows that by employing our methods, it is feasible to process a majority of events while generating fewer edges, compared to the method *radius*. Additionally, the uniform structure of the graph allows for the initialisation of an appropriate memory size.

Evaluation Results. The best mAP (mean Average Precision) results based on the methods used are presented in Table 2. The model achieving the highest accuracy values utilised normalisation to a value of 256, along with subsampling and the *radius* methods. Notably, this model outperformed the baseline model, which relied only on temporal normalisation, by achieving a 1.3% higher mAP value. Normalisation to a value of 64 significantly impacted object classification performance, resulting in poorer outcomes.

An analysis of graph directionality revealed that using a time-directed graph led to only a marginal 0.8% loss in mAP compared to an undirected graph. In contrast, the lowest performance was observed for the oppositely timed directed

---

[1] The N-Caltech101 dataset is an event-based version of the original Caltech101 dataset. It consists of 8246 samples distributed across 100 classes, with a maximum resolution of 240 × 180 pixels.

Table 2. Ablation study for accuracy for different methods.

Method	Normalisation	Pre-processing	Edge Generation	mAP
Base model [6]	Only time = 100	Random	Radius	53.47
Normalisation value	64	Random	Radius	43.73
	128	Random	Radius	51.99
	256	Random	Radius	54.77
Direction of edges	128	Random	NM - undirected	52.20
	128	Random	NM - time directed	51.40
	128	Random	NM - op. directed	50.23
Unique values	128	Unique	NM - time directed	**51.28**
	256	Unique	NM - time directed	**53.39**

Table 3. Comparison of different methods for object detection. * was not directly available and estimated based on the publication. Best results in **bold**.

Name	mAP	Parameters [M]
NvS-S [7]	34.6	0.9
AEGNN [10]	59.5	20.4
EAGR-N [5]	62.9	2.7*
EAGR-L [5]	**73.2**	30.7*
Our	53.39	**0.25**

graph, resulting in an almost 2% decrease compared to the undirected graph. Furthermore, applying all the optimisation methods, including normalisation to 256, only slightly affects the results with a 0.08% reduction in the mAP score.

Table 3 presents a comparative analysis of various GCN models in detection task on the N-Caltech101 dataset. Our work uses the model presented in [6]. Although its accuracy does not exceed that reported in [5], it significantly reduces the size of the model. The NvS-S model [7], the closest in size to ours, possesses three times as many parameters, but shows an accuracy lower by 18%. In contrast, the model [10], which heavily influenced the development of [6], achieves a 6.11% higher mAP at the cost of being 81 times larger in size.

The presented results demonstrate that, using the described optimisation methods, it is possible to construct a graph that enables effective object recognition with small accuracy losses. Also, our model achieves a good compromise between accuracy and size, making it well suited for use on embedded platforms.

4 Hardware Module Concept

We developed the optimisation methods for the representation of the event data graph (Sect. 3) taking into account the hardware constraints of FPGA platforms. In this context, the limitation of embedded memory usage and ensuring a deterministic number of operations are of key importance.

4.1 Hardware Module Requirements

From a functional point of view, the graph generation hardware module must meet the throughput and latency requirements to enable real-time event processing, even in highly dynamic scenes. The processing time of a single event affects the throughput of the system.

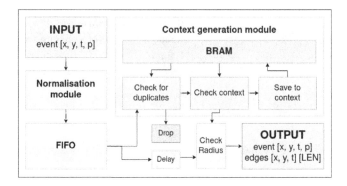

Fig. 1. The proposed hardware module scheme

The maximum number of events per second, after time quantisation, is $SIZE \times SIZE$ for a single time unit. However, a situation where the maximum number of events is present is highly unlikely. A realistic estimate of the required throughput was carried out based on an analysis of the number of events per time unit for the N-Caltech101 dataset. The maximum average number of events per 1 ms was determined to be approximately 3300, which means an average of ~3.3 events per 1 μs. The second major constraint on the implementation of the module is the limited amount of internal memory and logic resources of FPGAs. We address this by quantising input values and optimising memory footprint usage at the cost of a slight reduction in accuracy.

4.2 The Proposed Hardware Module

Based on the results of the GCN detection study and taking into account the specified system requirements, for the hardware module, we decided on a graph size of $SIZE = 256$, edges directed with time and a radius of $R = 3$. The module was implemented in the SystemVerilog language, simulated and synthesised using Xilinx Vivado tool for an AMD Xilinx ZCU 104 board with the Zynq UltraScale+ MPSoC chip (XCZU7EV-2FFVC1156). During testing, we assumed a module running synchronously with a clock frequency of at least 250 MHz.

Input, Normalisation and FIFO Modules. The module's input is a single event, described by the following signals: t (timestamp), x, y (coordinates) and

p (polarity). In addition, the *valid* flag makes it possible to determine whether an event has been received in a given clock.

The first operation performed on a received event is its scaling (for t, x and y values) and quantisation to $0 - SIZE$ values (Fig. 1, *normalisation* module). Normalisation performed immediately after receiving the event reduces the use of memory and logic resources. This operation is executed within a single clock, so it does not affect system throughput, only latency.

After normalisation, the event can be passed directly to the graph generation module. However, due to the asynchronous and sparse nature of the event data, we decided to add a FIFO queue here (Fig. 1, *FIFO*). In this way, events are accumulated in the queue in moments of increased scene dynamics. Such a mechanism allows efficient graph generation even for moments when the number of events per time significantly exceeds the system throughput. The length of the queue was set considering the availability of memory resources. After analysing the number of events per time, we assumed its size to be 1024 elements, which was sufficient for the test sequences used. The number of bits per memory cell was 25 (3 × 8-bit coordinates and 1-bit polarity) after normalisation.

Context Generation Module. The second element of the system is the analysis of the context of the event. As described in Sect. 3, we use a neighbour matrix (NM) that contains information about the most recent recorded events. To store its contents, we used a two-port BRAM memory of size $SIZE \times SIZE$, with one port for READ and WRITE operations and the other for READ only operations. To reduce the amount of data stored in the memory, we used the normalised event coordinates as the indexes of particular matrix cells, to which we wrote only the timestamp values. In this way, we could easily read the events in (x, y, t) format from the memory, decoding the coordinates x and y, while storing this data in just 8 bits (timestamp only).

For each quantised event read from the FIFO, the content of the memory cell corresponding to its coordinates (x, y) in the matrix is checked (Fig. 1, *Check for duplicates* module). If the timestamp value read from the NM is the same as in analysed event, it is dropped and another one is read from the FIFO queue.

Otherwise, a context of radius $R = 3$ pixels in form of a square with $2R + 1 \times 2R + 1$ shape is searched for potential edge candidates. The stored timestamps are read from the BRAM memory (Fig. 1, *Check context* module). Each READ operation during the processing of a single event takes 1 clock cycle (i.e., the latency is 1). To read the entire context (i.e., 48 memory cells), two ports of the BRAM memory are used, thus reducing the time of this operation twice – to 24 clock cycles for a single event. In practice, it may be sufficient to read fewer neighbours from the 7 × 7 context, using one of the approximate representations of a circle – then instead of the most time-consuming case of 48 candidates (for a square), 36 or 24 reads are sufficient, reducing the latency of the module and increasing the number of processable events per time. However, in our work, we developed the generation module for the worst-case scenario. It is worth noting that by increasing the width of the BRAM interface (storing multiple context candidates in one memory cell), the bandwidth of the system can be even higher.

After the entire context is read, a single WRITE operation to the BRAM memory takes place to update the timestamp value corresponding to the event being processed (Fig. 1, *Save to context* module). At the same time, a flag is generated indicating that the reading of the context has been completed.

Output: Edges Generation. The final step is to analyse the context and determine the output signals, which are pairs – each nondropped input event and a list of its edges. A simple delay line is used to pass the event (in the format (x, y, t, p)) to the module output. Each edge is described with the absolute values of vertices connected by an edge (x, y, t) stored on 24 bits. By normalising the time and dropping events with the same (x, y, t), an accurate description of a given vertex and described edge definition allows unambiguous identification of connected vertices. From such a generated edge representation, it is thus very easy to switch to the *pos* and *edges* structures described in Sect. 3.

Table 4. Hardware resource utilisation for the proposed edge generation module on an exemplary ZCU 104 platform for graph size of 256×256.

Resource type	Available	Used
LUT	230400	5612 (2.4)
Flip-Flop	460800	950 (0.2)
Block RAM	312	17 (5.5)
DSP	1728	189 (10.9)

To perform context-based edge generation, a neighbourhood condition of distance $\leq R$ (in each direction) is checked for each candidate (Fig. 1, *Check radius* module). We search for the neighbours inside the hemisphere $(x - x_c)^2 + (y - y_c)^2 + (t - t_c)^2 <= R^2$, where (x, y, t) are the coordinates of the vertex being processed and (x_c, y_c, t_c) are the coordinates of the candidate read from the context. Resulting pairs (vertex and its edge list) can then be written to external memory or passed to the GCN accelerator.

Theoretical Throughput. All logical operations in the proposed module take 1 clock cycle, so the processing of input events can be pipelined. However, the bottleneck of the solution is the communication with the BRAM memory, where the timestamp values are stored. A maximum of 26 clock cycles per single event are required: 25 for READ and 1 for WRITE operation, which was confirmed through simulation. Based on this information, we determined the theoretical throughput of the system to compare it with the specified requirements. For a 250 MHz clock, the system is able to accept a new event once every 104 ns, so over 9.6 events per 1 μs. The calculated theoretical throughput is almost 3 times higher than the requirements determined based on analysis of the N-Caltech101 dataset. Moreover, the use of FIFO module enables a high performance system for sequences with much higher temporary dynamics in the scene.

Hardware Utilisation. The proposed edge generation module was synthesised for an exemplary AMD Xilinx ZCU 104 board with the Zynq UltraScale+ MPSoC chip. The utilisation of hardware resources is presented in Table 4.

5   Conclusion

In this paper, we presented the concept of a hardware module that, based on the event stream, can generate a graph representation that can serve as an input to a graph convolutional network. In a series of experiments, we have shown that our proposed approach using directed (time-based) edges, subsampling and quantisation results in only a small loss of detection performance – 0.08% mAP for the N-Caltech dataset. The proposed module, in combination with a GCN accelerator and a SoC FPGA device, will allow to develop an embedded object detection system based on event data.

Our future objectives include developing a detection system and exploring several key areas: investigating diverse input data filtering and aggregation techniques, particularly those influenced by SNNs and tackling continuous event stream processing.

References

1. Abi-Karam, S., He, Y., Sarkar, R., Sathidevi, L., Qiao, Z., Hao, C.: Gengnn: a generic fpga framework for graph neural network acceleration. arXiv preprint arXiv:2201.08475 (2022). https://doi.org/10.48550/arXiv.2201.08475
2. Bi, Y., Chadha, A., Abbas, A., , Bourtsoulatze, E., Andreopoulos, Y.: Graph-based Object Classification for Neuromorphic Vision Sensing. In: 2019 IEEE International Conference on Computer Vision (ICCV). IEEE (2019). https://doi.org/10.1109/ICCV.2019.00058
3. Cordone, L., Miramond, B., Thierion, P.: Object detection with spiking neural networks on automotive event data. In: Proceedings of the IEEE International Joint Conference on Neural Networks (IJCNN) (July 2022). https://doi.org/10.1109/IJCNN55064.2022.9892618
4. Engelhardt, N., So, H.K.H.: GraVF-M: graph processing system generation for multi-FPGA platforms. ACM Trans. Reconfigurable Technol. Syst. (TRETS) **12**(4), 1–28 (2019). https://doi.org/10.1145/3357596
5. Gehrig, D., Scaramuzza, D.: Pushing the limits of asynchronous graph-based object detection with event cameras. arXiv preprint arXiv:2211.12324 (2022). https://doi.org/10.48550/arXiv.2211.12324
6. Jeziorek, K., Pinna, A., Kryjak, T.: Memory-efficient graph convolutional networks for object classification and detection with event cameras. In: 2023 Signal Processing: Algorithms, Architectures, Arrangements, and Applications (SPA), pp. 160–165. IEEE (2023). https://doi.org/10.23919/SPA59660.2023.10274464

7. Li, Y., et al.: Graph-based asynchronous event processing for rapid object recognition. In: 2021 IEEE/CVF International Conference on Computer Vision (ICCV), pp. 914–923 (2021). https://doi.org/10.1109/ICCV48922.2021.00097

8. Neu, M., et al.: Real-time Graph Building on FPGAs for Machine Learning Trigger Applications in Particle Physics. arXiv preprint arXiv:2307.07289 (2023). https://doi.org/10.48550/arXiv.2307.07289

9. Qi, C.R., Su, H., Mo, K., Guibas, L.J.: Pointnet: deep learning on point sets for 3d classification and segmentation. In: Proceedings of the IEEE Conference on Computer Vision and Pattern Recognition, pp. 652–660 (2017). https://doi.org/10.1109/CVPR.2017.16

10. Schaefer, S., Gehrig, D., Scaramuzza, D.: AEGNN: asynchronous event-based graph neural networks. In: Proceedings of the IEEE/CVF Conference on Computer Vision and Pattern Recognition (CVPR), pp. 12371–12381 (June 2022). https://doi.org/10.1109/CVPR52688.2022.01205

11. Sekikawa, Y., Hara, K., Saito, H.: EventNet: asynchronous recursive event processing. In: Proceedings of the IEEE/CVF Conference on Computer Vision and Pattern Recognition, pp. 3887–3896 (2019). https://doi.org/10.1109/CVPR.2019.00401

12. Tao, Z., Wu, C., Liang, Y., Wang, K., He, L.: LW-GCN: a lightweight FPGA-based graph convolutional network accelerator. ACM Trans. Reconfigurable Technol. Syst. 16(1), 1–19 (2022). https://doi.org/10.1145/3550075

Author Index

T. Dias and P. Busia (Eds.): DASIP 2024, LNCS 14622, p. 123, 2024.
https://doi.org/10.1007/978-3-031-62874-0

Printed in the United States
by Baker & Taylor Publisher Services